T0075238

Flexible Batteries

Flexible Batteries

Ye Zhang, Lie Wang, Yang Zhao, and
Huisheng Peng

CRC Press is an imprint of the
Taylor & Francis Group, an **informa** business

First edition published 2022
by CRC Press
6000 Broken Sound Parkway NW, Suite 300, Boca Raton, FL 33487-2742

and by CRC Press
4 Park Square, Milton Park, Abingdon, Oxon, OX14 4RN

CRC Press is an imprint of Taylor & Francis Group, LLC

ISBN: 978-1-032-22654-5 (hbk)
ISBN: 978-1-032-22673-6 (pbk)
ISBN: 978-1-003-27367-7 (ebk)

DOI: 10.1201/9781003273677

Typeset in Minion
by codeMantra

Contents

Preface

THE PAST DECADE HAS witnessed immense development in many new fields such as flexible, wearable, and medical electronics that may shape the future lifestyle of humankind. Conceivably, the devices used from these fields would be directly worn on or implanted in the human body and would work stably under complex deformations during use. As a result, there is an urgent need to develop corresponding batteries that are flexible and adaptable. However, traditional batteries that typically appear in rigid plate or bulk architectures cannot effectively meet the requirements for flexibility and adaptability. Therefore, the transformation of conventional batteries to flexible batteries, which are thin, lightweight, weavable, and even stretchable for integration, has become a milestone in the development of batteries.

This book summarizes the key advances in flexible batteries, a booming new direction in the energy storage field. We first introduce flexible lithium-ion batteries in both thin-film and fiber configurations as they are currently the most widely used batteries. Due to the safety concerns in organic electrolytes, flexible aqueous batteries such as aqueous lithium-ion, sodium-ion, and zinc-ion batteries are recently developed and discussed subsequently. Since the energy/power densities of conventional "rocking-chair" batteries are limited on the basis of their energy storage mechanisms, flexible metal-air batteries are recognized as primary choices for next generations due to their ultrahigh energy densities. Therefore, we take lithium-air and aluminum-air batteries as examples to explore their applications in flexible battery construction. We further summarize flexible batteries under the most challenging working conditions such as stretching and integrating flexible batteries with flexible energy harvesting devices, sensors, and supercapacitors. It covers both fundamental and application development that may be helpful to a variety of people with different backgrounds. For instance, how will flexible batteries differ from the well-studied planar batteries in charge transport? How can flexible batteries be used for many emerging fields including flexible, wearable, and biomedical electronics? The answers may be found in this book.

This book is also useful to scientists and engineers in the industry, where large-scale productions and applications of flexible batteries are gaining attention. Therefore, such a book may effectively bridge academics and industry. It is also intended for both professors and students of many departments including Materials Science and Engineering, Chemical Engineering, Chemistry, Physics, Energy Science, and even Biomedical Science. Flexible batteries may thus push the world ahead more rapidly when customers will have a very good understanding of such new technologies.

We sincerely acknowledge many graduate students who joined in writing this book: Chapter 1 was drafted by Yiding Jiao, Chapter 4 by Tingting Ye and Chang Jiang, Chapter 6 by Yiding Jiao, Chapter 7 by Qianming Li, Chapter 9 by Luhe Li, and Chapter 11 by Yiding Jiao. In particular, we would like to thank Yiding Jiao for revising the whole book at the final stage.

Ye Zhang, Lie Wang, Yang Zhao, Huisheng Peng

Authors

Professor Ye Zhang earned her PhD in Polymer Chemistry and Physics from Fudan University and carried out her postdoctoral research at Harvard University. She is now an associate professor of Nanjing University College of Engineering and Applied Sciences. Her research focuses on the development of soft electronics including batteries, sensors, and bioelectronic devices. She has published over 65 papers and authorized nine invention patents, two of which were transferred to industry. She was recognized with over ten academic awards including the IUPAC-Solvay International Award for Young Chemists, the Geneva International Invention Gold Award, and the CPCIF-Clariant Sustainability Youth Innovation Excellence Award.

Dr. Lie Wang is currently a postdoctoral scholar at the Nanjing University. He received his PhD in Macromolecular Chemistry and Physics from the Fudan University in 2020. His research focuses on the development of flexible energy and electronic devices and their wearable applications.

Professor Yang Zhao is currently an associate professor in Northwestern Polytechnical University. He received his PhD in Macromolecular Chemistry and Physics from Fudan University in 2020. His research interests are focused on the development of flexible and multi-functional energy storage devices with high safety.

Professor Huisheng Peng is currently a university professor at Fudan University. He received his BE in Polymer Materials from Donghua University in China in 1999, MS in Macromolecular Chemistry and Physics at Fudan University in China in 2003, and PhD in Chemical

Engineering from Tulane University in the USA in 2006. His interest focuses on the discovery and development of polymer fiber electronics. He has published over 300 papers and four books. Professor Peng has also obtained 79 licensed patents with 36 royally transferred to the industry.

CHAPTER 1

Introduction to Flexible Batteries

1.1 DEVELOPMENT OF BATTERIES

Since the first fire was set out, human beings have been dedicated to making full use of the various forms of energy available in nature. With the advancement in technologies, not only chemical energy from biomass and fossil fuels but also nuclear, solar, and wind energy have been widely explored. However, the acquisition and release of energy are normally transient, making the storage and transport impossible. Therefore, efforts have been made to store energy. In 1799, Italian scientist Alessandro Volta invented the first prototype of batteries, the voltaic pile, with copper, silver, and zinc metal piles and salt solution as electrolyte [1–3]. Since then, humankind has been exploring various kinds of batteries to store the fleeting energy in the form of chemical energy for use at different times and places in the form of electrical energy.

Despite the difference in materials and mechanisms, batteries are generally composed of two electrodes (a cathode and an anode), an electrolyte, and a separator. The energy is deposited through redox reactions of active materials in the electrodes. The realization of the redox reactions on separated electrodes is enabled by mass transfer in the electrolyte. To avoid short circuit, the separator is used to physically isolate the two electrodes. Therefore, during operation, ions in the electrolyte serve as charge carriers and migrate in the internal circuit while electrons transfer in the external circuit to create current.

DOI: 10.1201/9781003273677-1

Until now, lithium-ion batteries are undoubtedly the most successful commercialized batteries [4]. After careful investigation by John Goodenough, M. Stanley Whittingham, Rachid Yazami, and Koichi Mizushima during 1970s–1980s, lithium-ion batteries were first commercialized in 1990s by Sony Inc. [5]. Nowadays, the available lithium-ion batteries are generally composed of an anode based on graphitic carbon, a cathode based on Co, Mn, or $FePO_4$-based materials, and a Li-rich electrolyte based on organic solvent. The advantages of lithium-ion batteries are obvious. The energy density of lithium-ion batteries can reach 150–200 $Wh{\cdot}kg^{-1}$ with high working voltages of 3.3–4.2 V. Moreover, the cycle longevity is excellent for >1,000 cycles. However, the future development of lithium-ion batteries is hindered by intrinsic issues. The organic electrolyte suffers from severe safety hazards such as combustion and explosion. The shortage of lithium reserves also presages its pessimistic future. Despite the comparatively high energy density, efforts have also been made to further increase the energy density to power electronics with higher demands for energy.

Aqueous batteries are designed to replace the commercial lithium-ion batteries with organic electrolytes to fundamentally eradicate safety concerns. The intuitive attempt is the construction of lithium-ion batteries into aqueous ones using lithium aqueous electrolytes such as Li_2SO_4 aqueous solution. The as-fabricated aqueous lithium-ion batteries are endowed with working voltage of ~1.5 V and energy density of ~70 $Wh{\cdot}kg^{-1}$. However, the shortage of lithium reserves calls for aqueous batteries based on earth-abundant elements. Under this circumstance, aqueous sodium-ion batteries are developed with earth-abundant sodium element. Similar electrochemical performance as that of lithium-ion batteries can be obtained for sodium-ion batteries. Nevertheless, we are aware that designing alkali-metal-ion batteries into aqueous ones will dramatically diminish their energy density, thereby significantly restricting their application scenarios. Therefore, attempts have been made in two directions. Multivalent metal-ion aqueous batteries are developed, including magnesium-ion, calcium-ion, aluminum-ion, and zinc-ion batteries [6]. Due to the comparatively high capacity, low redox potential, cycle stability, and nontoxicity of zinc metals, aqueous zinc-ion batteries are preferred with energy density of > 200 $Wh{\cdot}kg^{-1}$ and cycle life from hundreds to thousands of cycles [7]. However, the ion intercalation working mechanism sets a limit to the energy density of these "rocking-chair" batteries. Therefore, efforts have also been made to replace the ion intercalation mechanism with direct redox reactions of metals. For example, lithium-air batteries were developed by realizing the direct reactions between lithium and O_2. A high theoretical energy density of ~3,500 $Wh{\cdot}kg^{-1}$ is expected, which is 5–10 times higher than those of conventional lithium-ion batteries. Still, the shortage of lithium reserves and the high reactivity of lithium call for other metal-air batteries [8]. Under this circumstance, aluminum-air batteries are developed with comparative theoretical energy density of 2,796 $Wh{\cdot}kg^{-1}$ and working voltage of 2.7 V [9].

1.2 FLEXIBLE BATTERIES

Since the development of batteries, a lot of efforts have been made to improve their electrochemical properties and design their shapes to fit different applications. For example, coin-type batteries have been developed for use in conventional watches, calculators, and thermometers, while cylinder, prismatic, and pouch-type batteries are used in smartphones and electric vehicles. However, all these configurations are bulky and rigid. The past decade has witnessed immense development in flexible electronic devices. Flexible and wearable electronic devices, such as Huawei Mate X flexible smartphone, Apple Watch, and Mi band, have been profoundly affecting the lifestyle of humankind. The latest report from the market research institute IDTechEx shows that the market will triple by 2026, compared with 10 years ago, and reach $150 billion. Moreover, flexible electronic devices have enlightened perspectives such as wearable sensors, implantable bioprobes, and interactive artificial intelligence [10]. However, the batteries appear in a rigid plate that cannot effectively meet the combined requirement on flexibility, light weight, miniaturization, and weaveability in modern electronics. Therefore, the transformation of conventional batteries to flexible batteries, which are thin, lightweight, weavable, and even stretchable for integration, has become a milestone in the development of batteries.

Generally, flexible batteries are fabricated into two configurations: two-dimensional thin films and one-dimensional fibers. The pioneering attempt to fabricate two-dimensional thin-film batteries was made by Oak Ridge National Laboratory using lithium phosphorous oxynitride solid-state electrolytes prepared by vacuum deposition in 1993 [11]. However, the high cost of fabrication and the time-consuming process have diminished its potential for real-time applications. Moreover, despite high conductivity and cost-effectiveness, conventional aluminum and copper current collectors are usually too stiff to undergo bending and folding deformations. Therefore, carbonaceous current collectors have been developed as viable substitutes for metal current collectors. Carbon cloth has been regarded as a promising current collector due to its structural stability, high conductivity, and commercial availability. The soft matrix enables not only improved ion/electron diffusion but also high mechanical strength against various deformations and volume changes. Graphene, which is a two-dimensional monolayer of carbon atoms with packed honeycomb lattices, exhibits large surface areas, thermal and chemical stability, high conductivity, and superior flexibility. These features facilitate the application of graphene as macroscopic self-standing electrodes. Carbon nanotube paper represents another competitive substrate for flexible batteries due to its intrinsic merits such as magnificent electrical and thermal conductivity, large surface area, outstanding chemical stability, light weight, and impressive mechanical properties. Therefore, carbon nanotubes can be made into membranes as substrates for active materials. Electrodes from carbon nanotube substrate deliver high cycling stability, good rate performance, and high gravimetric energy density. The free-standing electrodes from these carbonaceous substrates are further loaded with active materials and sandwich a solid-state electrolyte to form thin-film flexible batteries. These thin-film flexible batteries exhibit a certain degree of flexibility and are currently used in smart cards, radio frequency identification tags, implantable energy storage devices for neural stimulators, pacemakers, and defibrillators [12].

Although great efforts have been made, the flexibility of thin-film flexible batteries is still limited. In contrast, one-dimensional fiber batteries offer many intriguing advantages: (1) one-dimensional fiber batteries are compatible with the current textile industry and can be woven in textiles with multiple functions; (2) one-dimensional fiber batteries are permeable to air, solving the critical issue of thin-film batteries which are impermeable to air; and (3) one-dimensional fiber batteries exhibit superior flexibility, compatibility, and miniaturization capability, expanding their working scenarios. In 2013, Peng et al. designed a wire-shaped micro-lithium-ion battery by loading MnO_2 nanoparticles on aligned multi-walled carbon nanotube fibers with lithium wires. A specific volumetric capacity of 94.37 $mAh{\cdot}cm^{-3}$ and specific gravimetric capacity of 174.40 $mAh{\cdot}g^{-1}$ were achieved [13]. Since then, enormous efforts have been dedicated to developing more battery systems and widening their applications. To date, fiber lithium-ion batteries have been comprehensively developed while many novel battery systems such as sodium-ion batteries, zinc-ion batteries, and metal-air batteries have also been successfully transformed into fiber batteries [14]. Recently, metal-organic frameworks and liquid metals have been explored in fiber batteries, revealing their potential for next-generation fiber batteries [15].

Other than exploring novel materials and architectures, integration of flexible batteries with other functional components has also been a main theme in recent research. One direction is to integrate the flexible batteries with supercapacitors or energy conversion devices, which can simultaneously tackle the issues of low power density and self-power by obtaining energy from the environment. This strategy paves the way for higher energy efficiency and higher magnitude of multifunctionality. Another direction is to fabricate flexible batteries with other functional devices such as bioprobes and biosensors. Flexible batteries can power biosensors and form an integrated component for implantation. More efforts are currently made to design integrated systems with more functions and biocompatibility.

1.3 PERFORMANCE EVALUATION OF FLEXIBLE BATTERIES

The performance evaluation of flexible batteries can be divided into two categories: electrochemical property and flexibility. Electrochemical property generally follows the paradigm of conventional batteries, including charge/discharge curves, cycle longevity, battery capacities, rate capability, temperature characteristics, and power/energy densities. Charge/discharge curves signify the battery characteristics, including charge/discharge voltage plateaus, specific capacities, and operation stability represented from the continuity of the curves. Cycle life indicates the times a battery can be used under normal conditions, expressed in cycle numbers. Battery capacity refers to the overall charge stored in a charge/discharge cycle, expressed in mAh or Ah unit. For convenience in comparison, the battery capacity is usually normalized to gravimetric or volumetric specific capacities, which can be calculated by the following equations:

$$C_{\text{gravimetric}} = \frac{\int idt}{m}$$

$$C_{\text{volumetric}} = \frac{\int idt}{V}$$

where i is the constant current applied, t is the charge/discharge time, and m (V) is the mass (volume) of the active materials. Rate capability refers to capacity retention (%) with increasing charge/discharge rates, which relies on both the conductivity of fundamental components and reaction kinetics. Temperature characteristics indicate the reliability of batteries under abnormal working temperatures, including temperatures above and below the normal temperature range. Energy density is the amount of energy stored gravimetrically or volumetrically, in $Wh \cdot kg^{-1}$ or $Wh \cdot L^{-1}$, while power density is the amount of energy released per unit time, in $W \cdot kg^{-1}$ or $W \cdot L^{-1}$. The energy density can be calculated based on the following equation:

$$E_{\text{gravimetric}} = \frac{\int i \times Udt}{m}$$

$$E_{\text{volumetric}} = \frac{\int i \times Udt}{V}$$

where i is the constant current applied, t is the charge/discharge time, U is the potential as a function of t, and m (V) is the mass (volume) of the active material.

The power density can be calculated based on the following equation:

$$P = \frac{E}{T}$$

where E is either the gravimetric or volumetric energy density, and T is the total charge/discharge time.

In addition to electrochemical property, flexibility is another important parameter that requires evaluation. Flexibility of flexible batteries is commonly defined as the capability to maintain their electrochemical property when undergoing various deformations such as bending, twisting, and stretching. Among them, bending is the most applied working condition. The characterization of bending is generally based on two parameters:

bending angle (θ) and bending radius (r) of curvature. Some evaluation systems use a cylinder with a fixed bending radius as the substrate for flexible batteries. It is believed that the smaller r represents higher flexibility. However, the thickness and shape of the battery also significantly influence its flexibility. Therefore, the tensile strain calculated from the following equation is selected to describe the potential for flexible batteries in such application scenarios:

$$\varepsilon = h \times r^{-1}$$

where ε is the tensile strain, h is the thickness of the battery, and r is the bending radius.

Contrarily, the characterization of flexibility under twisting working condition is much cruder. A mostly accepted testing method is to stabilize the two ends of a flexible battery on a fixed actuator and a spinning actuator from the long side. Then, the spinning actuator spins within a fixed angle (e.g., ±15°) for certain twisting cycles (e.g., 1,000 cycles). The electrochemical property, which is normally the capacity retention, is recorded. The better flexibility under twisting deformation is signified with larger twisting angle, more twisting cycles, and higher capacity retention.

The harshest working condition for flexibility evaluation is stretching while no standard measurement has been well established. A typical measurement of this property is similar to that of twisting capability. The two ends from both long and short sides of a flexible battery are fixed to a static actuator and a tensile actuator. Subsequently, the tensile actuator stretches the flexible battery to a certain distance, normally in percentage of the length of the flexible battery (e.g., 300% of the length of the battery) for certain stretching cycles (e.g., 1,000 cycles). The electrochemical property, which is also the capacity retention, is recorded. The better flexibility under stretching deformation is represented by the longer stretching distance, more stretching cycles, and higher capacity retention.

REFERENCES

1. Sun, H. Zhang, Y. Zhang, J. Sun, X. Peng, H. 2017. Energy harvesting and storage in 1D devices. *Nature Reviews Materials* 2: 1–12.
2. Li, L. Wang, L. Ye, T. Peng, H. Zhang, Y. 2021. Stretchable energy storage devices based on carbon materials. *Small* 2005015.
3. Cecchini, R. Pelosi, G. 1992. Alessandro Volta and his battery. *IEEE Antennas and Propagation Magazine* 34: 30–37.

4. Tarascon, J. M. Armand, M. 2001. Issues and challenges facing rechargeable lithium batteries. *Nature* 414: 359–367.
5. Li, M. Lu, J. Chen, Z. Amine, K. 2018. 30 Years of lithium-ion batteries. *Advanced Materials* 30: 1800561.
6. Li, M. Lu, J. Ji, X. Li, Y. Shao, Y. Chen, Z. Zhong, C. Amine, K. 2020. Design strategies for nonaqueous multivalent-ion and monovalent-ion battery anodes. *Nature Reviews Materials* 5: 276–294.
7. Jia, X. Liu, C. Neale, Z. G. Yang, J. Cao, G. 2020. Active materials for aqueous zinc ion batteries: synthesis, crystal structure, morphology, and electrochemistry. *Chemical Reviews* 120: 7795–7866.
8. Grande, L. Paillard, E. Hassoun, J. Park, J.-B. Lee, Y.-J. Sun, Y.-K. Passerini, S. Scrosati, B. 2015. The lithium/air battery: still an emerging system or a practical reality? *Advanced Materials* 27: 784–800.
9. Liu, Y. Sun, Q. Li, W. Adair, K. R. Li, J. Sun, X. 2017. A comprehensive review on recent progress in aluminum-air batteries. *Green Energy & Environment* 2: 246–277.
10. Galos, J. Pattarakunnan, K. Best, A. S. Kyratzis, I. L. Wang, C.-H. Mouritz, A. P. 2021. Energy storage structural composites with integrated lithium-ion batteries: a review. *Advanced Materials Technologies* 6: 2001059.
11. Bates, J. B. Dudney, N. J. Gruzalski, G. R. Zuhr, R. A. Choudhury, A. Luck, C. F. Robertson, J. D. 1993. Fabrication and characterization of amorphous lithium electrolyte thin films and rechargeable thin-film batteries. *Journal of Power Sources* 43: 103–110.
12. Wang, C. Xia, K. Wang, H. Liang, X. Yin, Z. Zhang, Y. 2019. Advanced carbon for flexible and wearable electronics. *Advanced Materials* 31: 1801072.
13. Ren, J. Li, L. Chen, C. Chen, X. Cai, Z. Qiu, L. Wang, Y. Zhu, X. Peng, H. 2013. Twisting carbon nanotube fibers for both wire-shaped micro-supercapacitor and micro-battery. *Advanced Materials* 25: 1155–1159.
14. Dong, H. Li, J. Guo, J. Lai, F. Zhao, F. Jiao, Y. Brett, D. J. L. Liu, T. He, G. Parkin, I. P. 2021. Insights on flexible zinc-ion batteries from lab research to commercialization. *Advanced Materials* 33: 2007548.
15. Zhang, C. Zhu, J. Lin, H. Huang, W. 2018. Flexible fiber and fabric batteries. *Advanced Materials Technologies* 3: 1700302.

CHAPTER 2

Flexible Thin-Film Lithium-Ion Batteries

2.1 OVERVIEW OF LITHIUM-ION BATTERIES

Batteries have been invented for more than 200 years. A battery is a type of electrochemical device that can convert chemical energy into electrical energy to provide power for external devices. A battery usually consists of a cathode (or positive electrode) and an anode (or negative electrode), separated by an electrolyte with ionic conductive capability. The output voltage of the battery is derived from the potential difference between the two electrodes. According to rechargeability, batteries can be divided into two categories: primary batteries and secondary batteries. Secondary batteries are rechargeable batteries, which can be reused by recharging repeatedly. Cycling performance represents the capacity retention in charge-discharge cycles, which is a critical index for evaluating the quality of a battery. Obviously, secondary batteries are more environmentally friendly and energy saving than primary batteries and are currently more favored and widely used in various applications.

The first time lithium-ion batteries came into public sight was in 1990 when Sony Inc. announced a novel rechargeable battery that used $LiCoO_2$ as the cathode and a tailor-made carbonaceous material as the anode. Since its commercialization in the 1990s, lithium-ion batteries have been widely used as power supplies for various electronics, from mobile phones and laptops to electric vehicles [1,2]. In comparison with other secondary

DOI: 10.1201/9781003273677-2

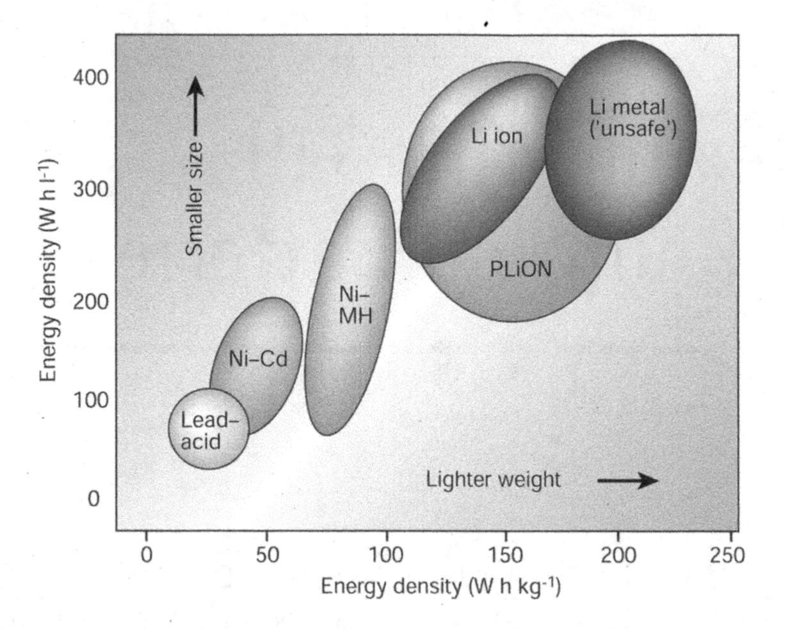

FIGURE 2.1 Gravimetric and volumetric energy densities of different rechargeable batteries. (Reproduced from Ref. [1] with permission of Nature Publishing Group.)

batteries (Figure 2.1), lithium-ion batteries are endowed with high energy density, high working voltage, and long cycling life, securing their leading position over competitors such as lead-acid, nickel-cadmium (Ni-Cd), and nickel-hydride (Ni-MH) batteries. Therefore, with the rapid development of wearable and implantable electronics, lithium-ion batteries are considered a viable option for such applications. However, the basic principles of lithium-ion batteries and the functions of different components should be first introduced in this section.

2.1.1 General Principle

The cathode materials for lithium-ion batteries are generally selected from lithium complexes with lithium intercalation capability. The anode material used for commercialized lithium-ion batteries was usually graphite. The cathode and anode materials are separated by a nonaqueous electrolyte, providing an output voltage due to their potential differences. The discharge capacity of the lithium-ion battery mainly derives from the reversible electrochemical redox reactions of active materials. Here, we take the $LiCoO_2$-graphite system as an example to introduce the

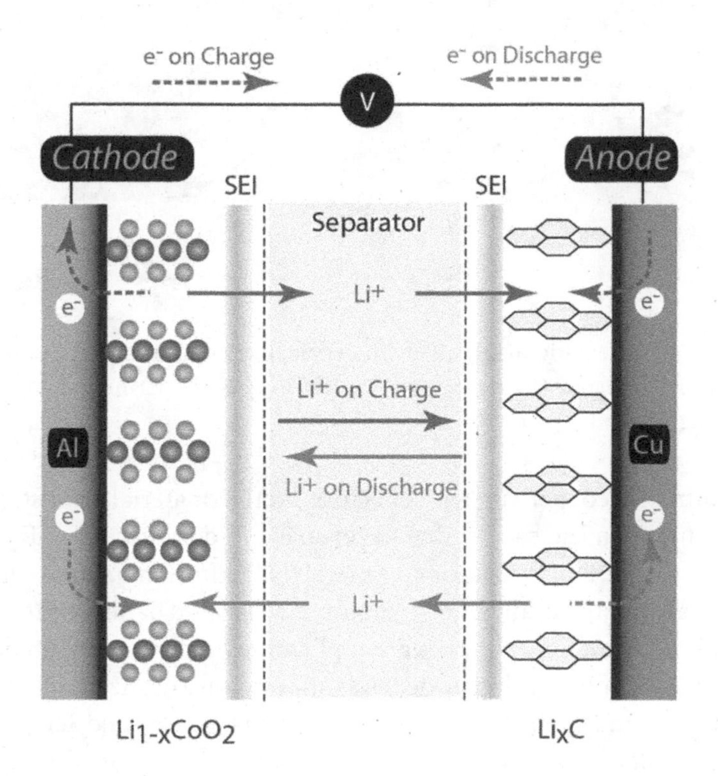

FIGURE 2.2 Schematic diagram of the charge and discharge process of lithium-ion batteries. (Reproduced from Ref. [3] with permission of the Royal Society of Chemistry.)

mechanism during the charge-discharge cycles. As shown in Figure 2.2, during the discharge process, lithium ions (Li^+) are removed from the anode, transferred *via* the electrolyte, and intercalated into the cathode. The whole process is reversed in the charge process, as illustrated in the following equations. Therefore, the lithium-ion battery is also called a "rocking-chair" battery.

$$\text{Cathode:} \quad Li_{1-x}CoO_2 + xLi^+ + xe^- \rightleftharpoons LiCoO_2$$

$$\text{Anode:} \quad xLiC_6 \rightleftharpoons xLi^+ + xe^- + xC_6$$

Despite cathode, anode, and electrolyte, separator and packaging materials are also required to establish lithium-ion batteries. Each component of a lithium-ion battery serves particular function for regular operation. The cathode and anode are generally fabricated by loading active materials on

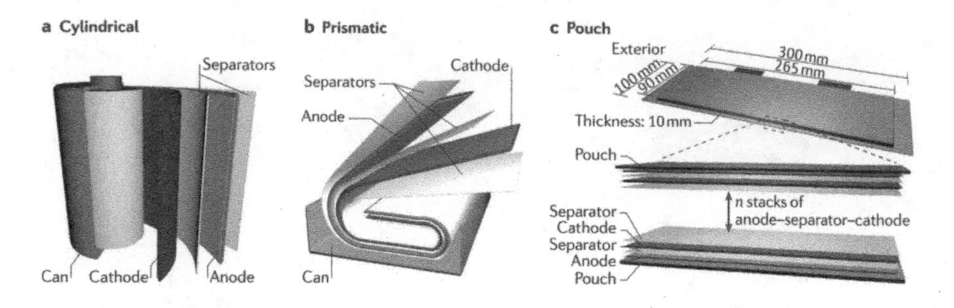

FIGURE 2.3 Schematic diagrams of the typical configurations of commercial lithium-ion batteries. (Reproduced from Ref. [4] with permission of Nature Publishing Group.)

aluminum and copper current collectors. Active materials can store and release energy in the battery due to reversible redox reactions. There are three typical configurations for commercial lithium-ion batteries, namely cylindrical, prismatic, and pouch (Figure 2.3). Several batteries are usually packed into a module in large-scale applications such as electric vehicles. Therefore, the module design depends mainly on the size and shape of the batteries and their interconnecting circuits, safety, and temperature control aspects. Cylindrical and prismatic batteries are generally rigid and inflexible while pouch batteries can achieve a certain degree of flexibility.

Due to the high theoretical capacity (3860 $mAh{\cdot}g^{-1}$), very negative standard redox potential (–3.04 V vs. standard hydrogen electrode), and low density (0.53 $g{\cdot}cm^{-3}$), lithium metal was used as the anode material in early attempts. However, significant parasitic reactions were brought by pristine Li anode, thus rendering limited cyclic longevity. Furthermore, lithium metal is prone to form dendrites, which penetrate the separator, leading to short circuit and safety hazards. Considering the adverse effects, lithium anode was replaced by graphitic carbon. It has been proven that Li^+ can be inserted into the crystal lattice of graphitic carbon, forming an intercalated compound (LiC_6), and the voltage profile presents a stable plateau at ~0.2 V vs. Li/Li^+ during lithiation/delithiation. The lithiated compound, LiC_6, determines that the theoretical lithium storage capacity is 372 $mAh{\cdot}g^{-1}$ for graphitic carbon.

Upon the intercalation of Li^+, graphite experiences several stages of phase transition spanning from LiC_{27}, LiC_{24}, and LiC_{12} to the final product LiC_6 [5]. In the first charge cycle in the polar aprotic electrolyte, a passivation film is apt to form at the surface of the anode, derived from the decomposition of electrolyte, which on the one hand protects the graphite from deterioration, but, on the other hand, causes initial irreversible capacity. Hence, excessive lithium has to be stored in electrodes to satisfy the consumption from the passive film formation.

For cathode materials, layered compounds where Li^+ are readily intercalated with stable structures are preferred, as exemplified by transition metal sulfides and $LiCoO_2$. Titanium disulfide (TiS_2) is one of the most appealing layered dichalcogenides applicable as energy storage materials. When compounded with Li^+, lithiated compounds of Li_xTiS_2 are exempt from the rearrangement and reconstruction of lattice, which contributes to their remarkable reversibility. In 1980, Goodenough *et al.* invented a lithium-ion battery using $LiCoO_2$ as the cathode material. $LiCoO_2$ is endowed with a similar layered structure with TiS_2 but exhibits different lithiation behaviors involving several phase transitions. Therefore, only half of the lithium in the compound can be exploited, leading to a relatively low theoretical capacity of 140 $mAh{\cdot}g^{-1}$. In recent years, more cathode materials like $LiNi_{1/3}Mn_{1/3}Co_{1/3}O_2$, $LiMn_2O_4$, $LiNiO_2$, $LiNiPO_4$, and $LiFePO_4$ have been developed.

For electrolyte selection, aqueous electrolytes are first excluded because the working voltage of lithium-ion batteries is far above the decomposition voltage of water (1.23 V vs. standard hydrogen electrode). Therefore, electrolytes with a lithium salt dissolved in aprotic solvents are effective in lithium-ion batteries, with wide electrochemical windows and good compatibility. Since the electrodes in lithium-ion batteries are comprised of active materials, conducting additive, and polymer binder, the liquid electrolytes must infiltrate into the porous electrodes and transfer Li^+ smoothly at the interfaces between the liquid and solid phases [6]. A separator is placed between cathode and anode to prevent short circuit between electrodes, which helps isolate the electronic flow and enables free ionic transport. In the following sections, we will introduce the cathode, anode, electrolyte, and separator in detail.

2.1.2 Cathodes

To achieve lithium-ion batteries with high output voltage and high energy density, the lithium exchange reaction of cathode materials must occur at a high potential relative to lithium [7]. Furthermore, the cathode materials must be insoluble in the electrolyte. Many cathode materials have been developed, many of which are commercially available. The output voltage

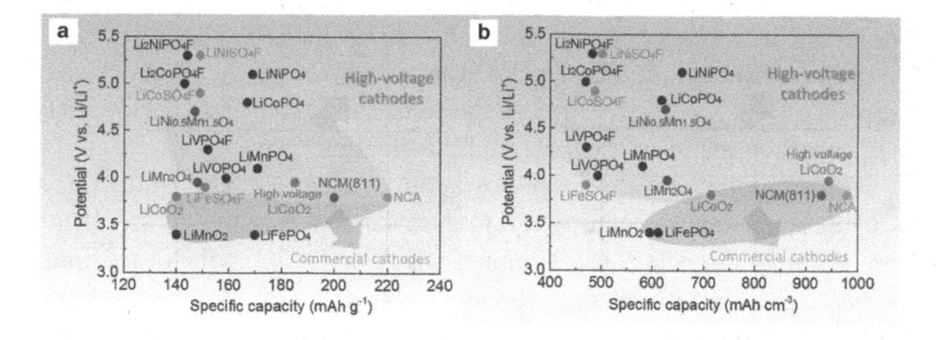

FIGURE 2.4 Gravimetric (a) and (b) volumetric capacities of the cathode materials of lithium-ion batteries. (Reproduced from Ref. [9] with permission of the Royal Society of Chemistry.)

and capacity determine the energy of a lithium-ion battery. A high voltage and capacity and lightweight promise a high energy density, which is badly desired and vigorously pursued [8]. Therefore, cathode materials play a pivotal role in increasing the energy densities of lithium-ion batteries. The voltage and capacity characteristics of common cathode materials (e.g., $LiCoO_2$, $LiNi_{1/3}Mn_{1/3}Co_{1/3}O_2$, $LiMn_2O_4$, and $LiFePO_4$) are summarized in Figure 2.4.

The most used positive electrode material is $LiCoO_2$, a layered structure with a high capacity of 274 mAh·g^{-1}. However, only half of the lithium in $LiCoO_2$ is cyclically available, corresponding to the charge ceiling potential of 4.2 V (vs. Li/Li^+). Moreover, the material structure inclines to collapse due to a significant electrostatic repulsion among neighboring transition metal polyhedral units. As a result, the reversible capacity of $LiCoO_2$ is typically limited to around 140 mAh·g^{-1}. Therefore, strategies have been designed to partially replace a fraction of the cobalt in this phase with nickel/manganese to preserve the structure against decomposition at high potential. The stoichiometric formula of the resulting cathode material is $LiCo_{1/3}Ni_{1/3}Mn_{1/3}O_2$, which promises a reversible capacity of 190 mAh·g^{-1} [10].

Compared with $LiCoO_2$, $LiMn_2O_4$ with a spinel structure proves to be much more robust during the removal of lithium. $LiMn_2O_4$ is much cheaper than $LiCoO_2$ and can reversibly remove and restore utmost 95% of the theoretical capacity. The combination of the three-dimensional spinel lattice with the chemical stability of the Mn^{+3}/Mn^{+4} couple promises good safety and high-power performance for a cathode material. Unfortunately, $LiMn_2O_4$ also suffers from stability problems such as self-discharge when used in conventional liquid electrolytes where the trace amounts of hydrogen fluoride (HF) will corrode the cathode materials, especially at elevated temperatures. $LiMn_2O_4$ with a spinel structure is one of the fastest cathode materials due to its unique three-dimensional structure, allowing the multidirectional diffusion of Li^+ to deliver 80 $mAh{\cdot}g^{-1}$ at high discharge rates [11,12].

As another cathode material, $LiFePO_4$ with an olivine structure has many advantages, including low cost, high capacity, no toxicity, environmental friendliness, and chemical and thermal stability. $LiFePO_4$ has a theoretical reversible capacity of 170 $mAh{\cdot}g^{-1}$ with a discharge potential of around 3.45 V (vs. Li/Li^+). Unfortunately, $LiFePO_4$ suffers from its insulating nature (10^{-9} $S{\cdot}cm^{-1}$), leading to slow Li^+ diffusion ($<10^{-14} cm^2{\cdot}s^{-1}$), high impedance, and inferior rate capability, impeding them toward commercial applications [13]. In consequence, in order to improve the electrochemical performance of $LiFePO_4$ cathode materials, various efforts have been devoted to increasing the electrical conductivity and decreasing the particle size of $LiFePO_4$. For example, the performances of $LiFePO_4$ can be improved by structural or morphological modifications such as mechano-chemical activation, conducting additives coating, and supervalent metal ion doping [14,15].

2.1.3 Anodes

Anode materials have an essential influence on the safety, energy density, and cycle life of lithium-ion batteries [16]. Lithium metal was used as the anode material in the early study. However, it was found that lithium metal brought about unwanted parasitic reactions on the lithium surface, thus rendering the battery limited cyclic longevity. Worse, lithium metal is prone to form lithium dendrites that can pierce the separator and cause internal short circuit and substantial safety risks. Considering the safety concerns, finding suitable substitutes became a primary issue in the research of anode materials.

In 1971, Dey demonstrated the possibility of electrochemical formation of lithium alloys in organic electrolytes [17]. Since then, alloys with aluminum, silicon, and zinc have been under investigation as alternative anode materials. The lithium storage capacity of alloys degrades quickly after several charge-discharge cycles arising from the significant volume change, which causes internal stress in the alloy crystal lattice, bringing cracking and pulverization problems of alloy particles. In 1991, Sony Inc.

succeeded in discovering the highly reversible, low voltage carbonaceous material as the anode and commercialized the "rocking-chair" lithium-ion battery using $LiCoO_2$ as the active cathode material and graphitic carbon material as the anode material. The growth of irregular dendritic lithium is prevented by replacing the lithium metal with lithiated carbon, which essentially improves the compatibility and safety for practical applications. The voltage and capacity characteristics of common anode materials are summarized in Figure 2.5.

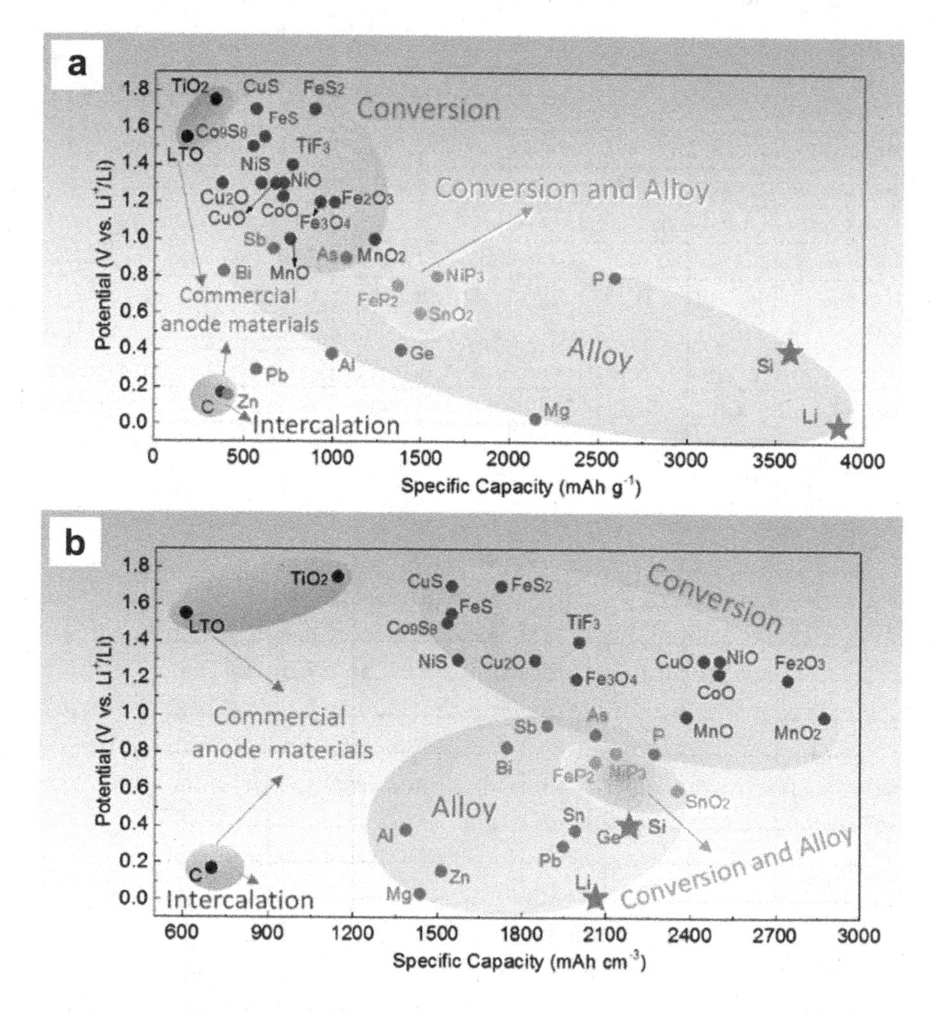

FIGURE 2.5 (a) Gravimetric and (b) volumetric capacities of anode materials of lithium-ion batteries. (Reproduced from Ref. [9] with permission of the Royal Society of Chemistry.)

The layered carbon was found accessible for various species, and the capability for intercalation promoted the application of graphite as an ion host in rechargeable batteries. Graphite has a theoretical specific capacity of 372 mAh·g^{-1} for lithium storage under normal pressure [18]. For a specific cathode material, with its capacity (C_C) ranging from 140 to 200 mAh·g^{-1}, there is a simple relationship between the total capacity of the lithium-ion battery (based on all active materials) and the anode capacity (C_A):

$$C_{\text{battery}} = \frac{C_c}{1+\frac{C_c}{C_A}} = \frac{C_A C_C}{C_A + C_C}$$

According to the above equation, for a given C_C, the total specific capacity of the battery does not follow a linear relation with C_A. A noticeable improvement in lithium-ion battery can be achieved if the carbonaceous anode is replaced with an anode having a capacity in the order of 1,000 mAh·g^{-1} (Figure 2.6) [19].

The potential of the alloy anode vs. Li/Li$^+$ should be close to 0 V, and the electrochemical reaction at the anode site does not necessitate a lithium intercalation process whose capacity is restricted by the crystal structure.

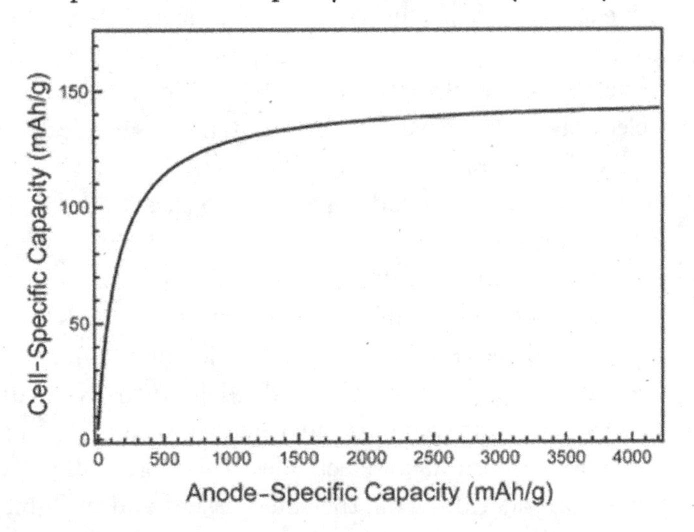

FIGURE 2.6 The relation between the total capacity of a lithium-ion battery with the anode capacity. The battery comprises a Si anode and $LiCoO_2$ cathode (the theoretical capacity of $LiCoO_2$ is 147 mAh·g^{-1}). (Reproduced from Ref. [19] with permission of the Royal Society of Chemistry.)

It has been demonstrated that Li^+ can react with the metal to form an alloy (Li_xM), which can accommodate and deliver more Li^+. Numerous metals such as Si, Al, Sn, Sb, Ge, Pb, and Ag can alloy electrochemically with lithium when polarized to a sufficiently negative potential in the electrolyte. For example, Si has a high theoretical capacity of 4,200 $mAh·g^{-1}$. The formation and decomposition of Li_xM provide a high theoretical capacity for the anode. However, most high-capacity anodes based on the metals above suffer from the short cycle life arising from the significant volume expansion and electrolyte depletion [16].

2.1.4 Electrolytes

An electrolyte for lithium-ion batteries typically contains ion-conducting components composited of functional or structural components such as additives and structural enhancements [20]. The primary function of the electrolyte is to transport Li^+ between two electrodes while impeding electrons efficiently. The desired electrolyte properties are different for energy storage devices with various structures, types, and working mechanisms [21]. The electrolytes for lithium-ion batteries can be divided into three types: liquid, solid, and gel electrolytes. The merits of liquid electrolytes are the high ionic conductivity and capability to form excellent and stable contact with electrodes. With the rapid development of flexible and wearable electronics, solid or gel electrolytes are preferred as they are safer and suitable for flexible and multifunctional lithium-ion batteries.

Liquid electrolytes have played an essential role in electrochemical energy storage for several decades due to their high ionic conductivities (10^{-3} to 10^{-2} $S·cm^{-1}$) and good contacts with electrodes [6]. The commonly used lithium salts for liquid electrolytes are $LiPF_6$, $LiClO_4$, $LiBF_4$, and $LiAsF_6$, which should be fully soluble or dissociated in nonaqueous solvents without reacting with anions. However, the use of liquid electrolytes has brought leakage risks or even combustion of organic electrolytes. Besides, the inevitable growth of lithium dendrites usually occurs in liquid solutions caused by uneven currents when charged in porous separators, especially for lithium metal electrodes. Therefore, solid electrolytes with no liquid solvents can tackle the safety issues and prohibit lithium dendrite growth.

Solid electrolytes are particularly promising for flexible lithium-ion batteries due to their ensured safety and excellent mechanical properties, which can be cut and shaped randomly [22]. There are two main types of solid electrolytes: inorganic electrolytes and polymer electrolytes. Inorganic electrolytes such as lithium phosphorous oxynitride (LiPON) thin films are being widely used. Solid polymer electrolytes were pioneered by Wright *et al.* in 1973. Various solid polymer electrolytes such as poly(ethylene oxide) (PEO)-based electrolytes have been studied deeply due to their high solubility for lithium salts [23]. PEO is a typical semi-crystalline polymer with poor electrical conductivity. However, it can form various complexes with lithium salts because of the coupling effects between Li^+ and oxygen atoms on the PEO chains, which improve the dissolution of lithium salts.

Copolymers with conductive blocks (usually PEO blocks) are promising candidates for high-performance solid polymer electrolytes. The self-assembled microstructures of copolymer electrolytes offer a good balance between ion conductivity and mechanical performance. As electrolytes, solid polymer electrolytes are free of safety issues with the merit of lightweight and high mechanical strength. However, solid electrolytes also face some critical problems, including low ionic conductivity and poor interfacial properties [24]. It is highly desirable to use high ionic conductivity at room temperature, good shear modulus, and high thermal and electrochemical stability for improving the performance of solid polymer electrolytes.

Gel electrolytes combine the merits of liquid and solid electrolytes, i.e., high ionic conductivity and good interfacial properties such as liquid electrolytes and good mechanical properties like solid electrolytes, which can function as electrolytes and separators [25,26]. The flexibility and elasticity of gel polymer electrolytes are also beneficial for tolerating the volume change of electrode materials and the formed dendrites during charge and discharge processes. Similarly, the gel electrolytes for lithium-ion batteries are typically composed of gel polymer as host, lithium salt, and organic solvent. Polar polymers such as PEO, polyacrylonitrile (PAN), polymethyl methacrylate (PMMA), and poly vinylidenefluoride-hexafluoropropylene (PVDF-HFP) are the commonly used polymer hosts for gel electrolytes owing to their excellent affinity with liquid components. Many related studies on gel electrolytes focus on enhancing the ionic conductivity, mechanical strength, formability, and contact with electrodes by optimizing their compositions and structures [27]. Significantly, gel electrolytes are promising for portable and flexible electronics and endow flexible lithium-ion batteries with higher flexibility and safety.

2.2 FLEXIBLE THIN-FILM LITHIUM-ION BATTERIES

Wearable electronic products such as the Apple Watch, Samsung bracelets, and smart clothes are emerging in the mainstream and represent promising future lifestyles [28–30]. Conceivably, these products would be directly worn on the human body and work stably under complex deformations, including bending, folding, and even stretching. The rapid advancement of flexible and wearable electronic products strongly demands interconnected power systems that are flexible, miniaturized, and wearable. However, most commercially available lithium-ion batteries are rigid and cannot effectively meet the requirements above, which has become a bottleneck for the further development of wearable electronics. Therefore, it is crucial to develop flexible lithium-ion batteries for wearable electronics [31,32].

Flexible electrode materials are critical components to construct flexible lithium-ion batteries. Compared with traditional metallic materials, carbon materials demonstrate lower mass density and higher intrinsic flexibility derived from microscopic structures. Carbon materials (such as carbon nanotubes (CNTs) and graphene) usually exhibit outstanding thermal and chemical stability, high electrical conductivity, and superior mechanical properties (Table 2.1). The high chemical stability and wide potential window of carbon materials also preclude corrosions and benefit the cyclic reversibility. Furthermore, great reaction sites and high specific surface areas in carbon nanomaterials facilitate the solid loading of active materials on them, which is beneficial for good electrochemical performance [33]. Therefore, carbon materials are often used in flexible electrodes for flexible thin-film lithium-ion batteries [34–36].

According to the used electrode materials, the flexible thin-film lithium-ion batteries can be divided into different types based on CNTs, graphene, carbon cloth, and other conductive materials, and they are detailed below.

2.2.1 Carbon Nanotube-Based Flexible Lithium-Ion Batteries

CNTs have a high specific surface area (1,600 $m^2 \cdot g^{-1}$), excellent electrical conductivity (10^3 $S \cdot cm^{-1}$), and suitable mechanical properties, which have been widely investigated and used in energy storage devices [3,44–46]. Different kinds of methods have been developed to build CNT-based electrodes for flexible lithium-ion batteries such as blade coating, drop-casting, vacuum filtration, solvothermal method, and chemical vapor deposition [47].

TABLE 2.1 Physical Properties of Carbon Nanomaterials [37–43]

Material	Method	Tensile Strength (MPa)	Young's Modulus (GPa)	Conductivity ($S \cdot cm^{-1}$)	Specific Surface Area ($m^2 \cdot m^{-3}$)
CNT	Drawing CNT array	600	74	10^3	–
Reduced giant graphene oxide	Wet spinning	360.1±12.7	12.8±0.8	3.2×10^4	–
Reduced graphene oxide and silver nanowire	Wet spinning	305		4.1×10^4	–
MWCNT	Slicing	–	–	Up to 10^3	–
Graphene	Wet spinning	69	–	1.2×10^3	–
Graphene	Hydrothermal	–	–	5×10^{-3}	–
CNT	Template directed	–	–	~1	~10^4

To realize a flexible thin-film lithium-ion battery, a thin and flexible lithium-ion battery was designed using paper as separator and free-standing CNT thin films as current collectors for cathode and anode [48]. The free-standing thin films of the cathode and anode were prepared through a coating and peeling process. By a simple lamination process, the double layer films of cathode and anode were laminated on the two sides of the paper to obtain the lithium-ion paper battery (Figure 2.7a). Before encapsulation for measurement, the obtained paper battery had a laminate structure with flexibility (Figure 2.7b and c). The lithium-ion paper battery was proven to light up a red LED continuously for 10 min without fading (Figure 2.7d). The paper battery could typically work under bending, which showed very high flexibility due to the high softness and low thickness of CNT-based film electrodes (Figure 2.7e).

The first cycle galvanostatic charging/discharging curves of the LTO-LCO paper battery with a thickness of ~10 μm were shown in Figure 2.7f. The coulombic efficiency of the first cycle was 85%, which was relatively low compared with the typical $LiCoO_2$-$Li_4Ti_5O_{12}$ (LCO-LTO) batteries. As shown in Figure 2.7g, the coulombic efficiency of lithium-ion paper battery increased to 94%–97% after the first cycle. After 20 cycles, 93% of the initial capacity could be maintained. Furthermore, the full paper battery

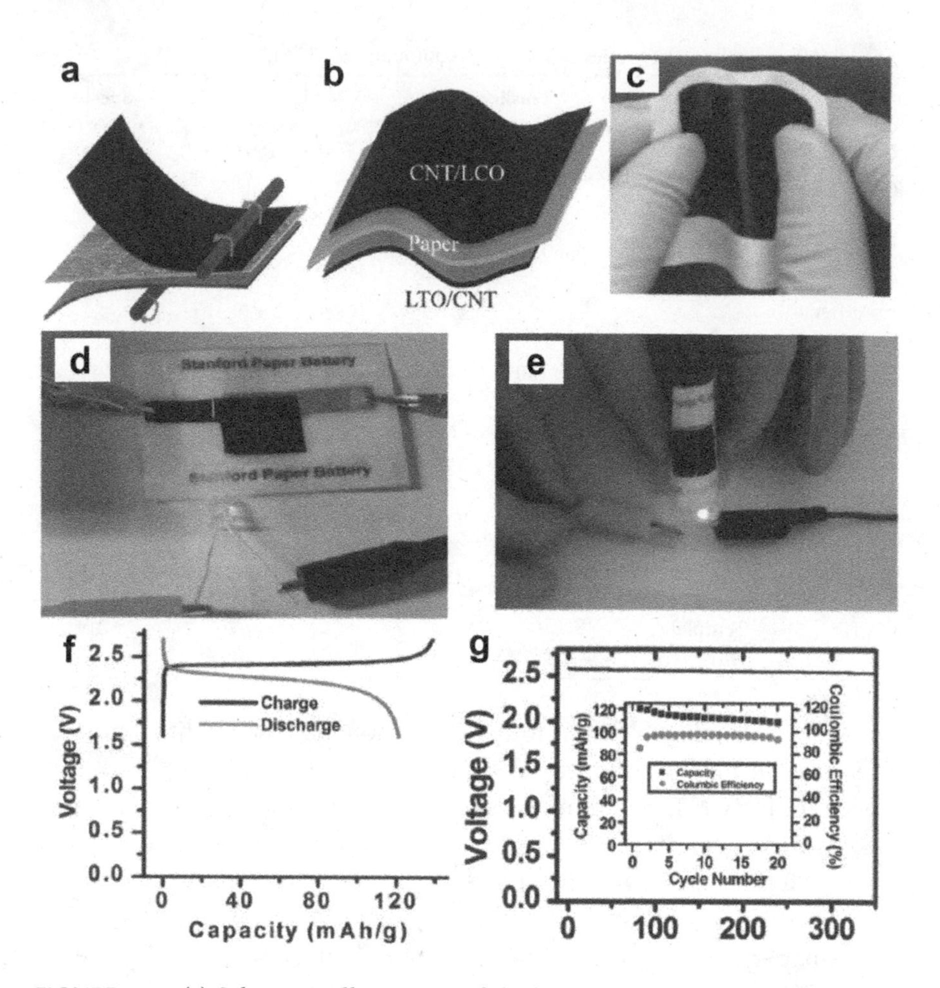

FIGURE 2.7 (a) Schematic illustration of the lamination process. (b) Schematic structure of the final paper lithium-ion battery device. (c) Photograph of the Lithium-ion paper battery before encapsulation for measurement. (d) Photograph of a PDMS-encapsulated lithium-ion paper battery lighting a red LED. (e) Photograph of flexible lithium-ion paper batteries lighting an LED device. (f) Galvanostatic charging/discharging curves of the LTO-LCO paper battery. (g) Self-discharge behavior of a full battery after being charged to 2.6 V. The inset is the cycling performance of the LTO-LCO full battery. (Reproduced from Ref. [48] with permission of the American Chemical Society.)

showed a 5.4 mV small voltage drop after 350 h, which demonstrated good self-discharge performance and was crucial for practical applications.

A Fe_2O_3/single-walled CNT (Fe_2O_3/SWCNT) film was designed with high Fe_2O_3 loading for the flexible anode of lithium-ion battery [49].

After an oxidizing process of the flow-assembly Fe/SWCNT membrane, uniform Fe_2O_3 nanoparticles (sizes of 5–10 nm) were homogeneously anchored on SWCNTs (Figure 2.8a). The Fe/SWCNT membrane (diameter of ~45 mm, Figure 2.8b) was used as the product collector during growing, and the membrane's thickness could be adjusted. Fe nanoparticles were then anchored on the membrane of interwoven SWCNT network (Figure 2.8c). After the oxidizing process at 390°C, the membrane was still flexible (Figure 2.8d), and the Fe nanoparticles were oxidized into iron oxide nanoparticles (Figure 2.8e). The Fe_2O_3/SWCNT membrane electrode demonstrated initial discharge and charge capacities of 2,097 and 1,243 $mAh \cdot g^{-1}$ (Figure 2.8f), which were much larger for the SWCNT membrane (1,651 and 467 $mAh \cdot g^{-1}$, Figure 2.8g). In the following cycles, the charge and discharge capacity was well maintained, which showed good cycling performance of Fe_2O_3/SWCNT and SWCNT membrane electrodes.

A 3D hierarchical SnO_2 nanowire (NW) core-amorphous silicon shell on free-standing CNT paper (SnO_2@a-Si/CNT paper, Figure 2.9a) was also made as a flexible anode for lithium-ion battery [50]. The SnO_2 NWs demonstrated 40–120 nm diameters with typical lengths of several hundred micrometers (Figure 2.9b). The CNT paper and SnO_2@a-Si NWs formed a 3D network consisting of entangled nanotubes and highly dense nanowires (Figure 2.9c). The first discharge capacity of the SnO_2@a-Si/CNT electrode was 3,020 $mAh \cdot g^{-1}$ (Figure 2.9d). The SnO_2@a-Si/CNT electrode demonstrated better lithium storage stability than SnO_2/CNT electrode (Figure 2.9e). The rate capability of the electrode was shown in Figure 2.9f. After 25 rate cycles, the SnO_2@a-Si/CNT electrode showed high structure stability (Figure 2.9g and h).

2.2.2 Graphene-Based Flexible Lithium-Ion Batteries

Graphene is a 2D sp^2-bonded carbon sheet with a thickness of one atom. Compared with CNTs, graphene has a higher theoretical surface area (2,630 $m^2 g^{-1}$), providing numerous active sites for electrochemical reactions. Besides, graphene shows unique properties with simultaneously high mechanical flexibility, electrical conductivity, thermal, and chemical stability. These outstanding properties make graphene promising for flexible energy storage devices. The functionalization of graphene is relatively easy, which allows the preparation of graphene-based hybrid electrodes with good performance.

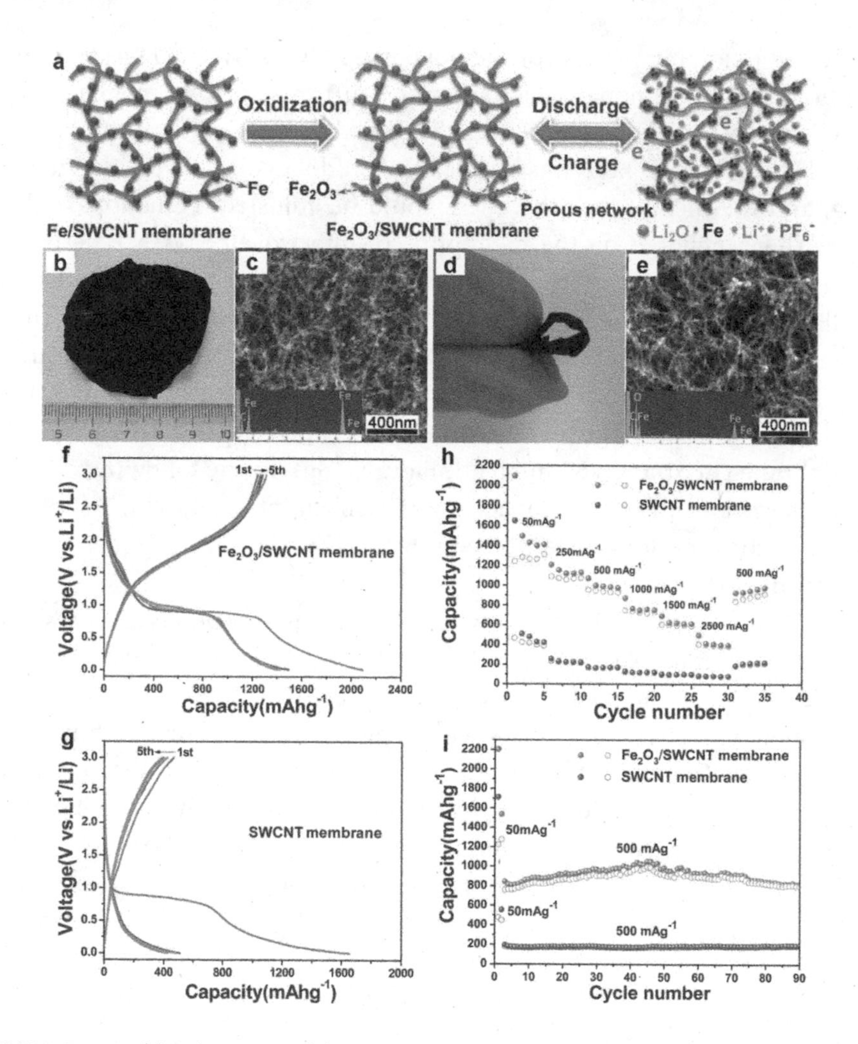

FIGURE 2.8 (a) Schematic of the preparation process and the structure of Fe_2O_3/SWCNT membrane. (b) Photograph of a Fe/SWCNT membrane (diameter of ~45 mm). (c) SEM image of the Fe/SWCNT membrane and the energy-dispersive X-ray spectroscopy (EDS) spectrum. (d) Photograph of the flexible membrane obtained after the oxidizing process. (e) SEM image of the Fe_2O_3/SWCNT membrane and the EDS spectrum. (f) Galvanostatic charge-discharge curves of the flexible Fe_2O_3/SWCNT membrane. (g) Galvanostatic charge-discharge curves of the SWCNT membrane. (h) Rate performance of the Fe_2O_3/SWCNT and SWCNT membranes at different current densities. Solid symbols, discharge and hollow symbols, charge. (i) Cycling performance of the Fe_2O_3/SWCNT and SWCNT membranes at a current density of 500 mA·g^{-1}. Solid symbols, discharge and hollow symbols, charge. (Reproduced from Ref. [49] by permission of the Royal Society of Chemistry.)

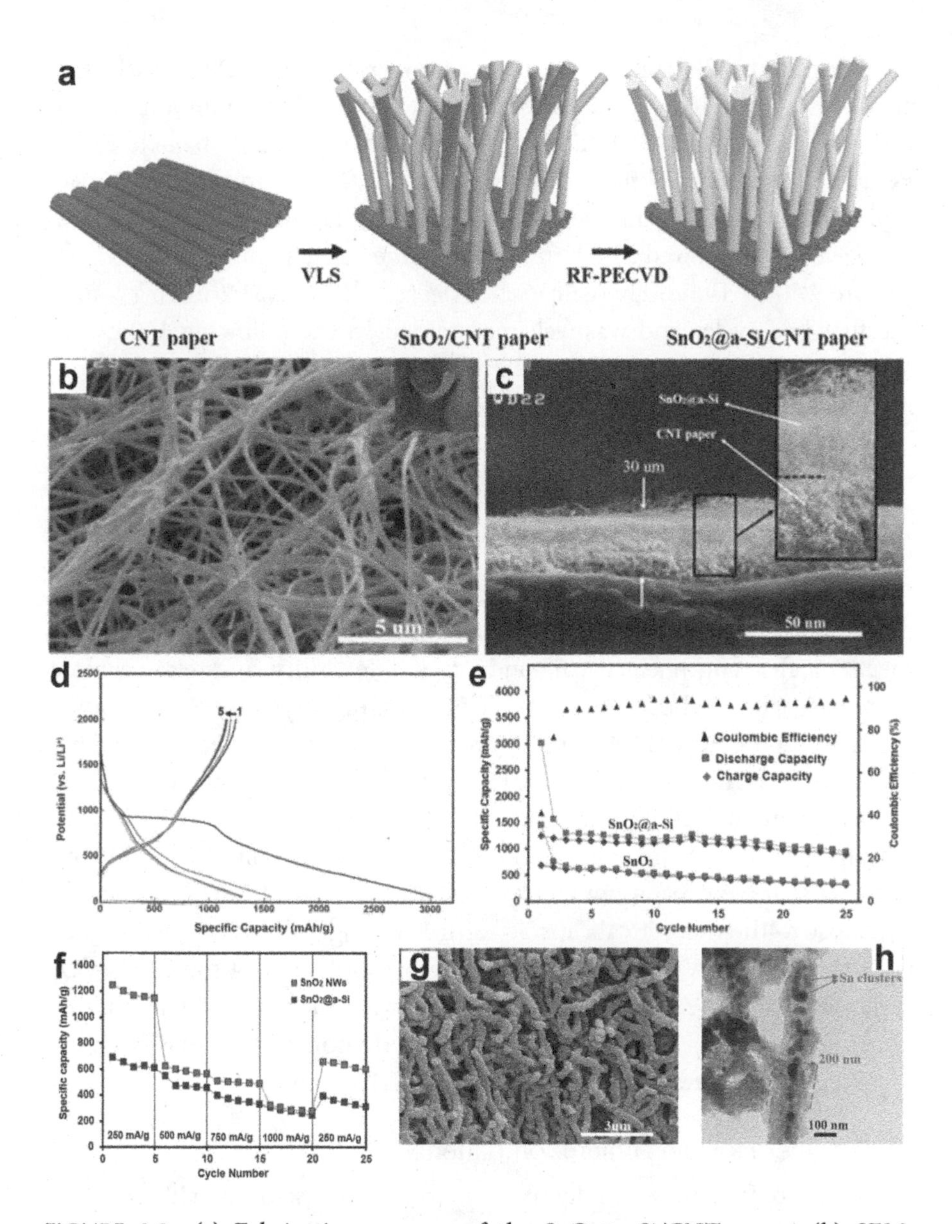

FIGURE 2.9 (a) Fabrication process of the SnO_2@a-Si/CNT paper. (b) SEM image of the SnO_2@a-Si/CNT paper. (c) Cross-sectional SEM image of the SnO_2@a-Si/CNT paper. (d) Galvanostatic charge-discharge curves of the SnO_2@a-Si/CNT electrode. (e) Cycling performance of the SnO_2/CNT and SnO_2@a-Si/CNT electrodes. (f) The rate capability of the SnO_2/CNT and SnO_2@a-Si/CNT electrodes. (g, h) SEM and TEM images of the SnO_2@a-Si/CNT anode after 25 cycles. (Reproduced from Ref. [50] with permission of IOP Publishing Ltd.)

A flexible binder-free three-dimensional graphene/Fe_2O_3 (3DG/Fe_2O_3) aerogel (Figure 2.10a) was used as anode for a flexible lithium-ion battery [51]. The as-prepared 3DG/Fe_2O_3 could be compressed to flexible 3DG/Fe_2O_3 film with a thickness of ~50 μm (Figure 2.10b), and the Fe_2O_3 was well-encapsulated within the 3DG/Fe_2O_3 film (Figure 2.10c). The 3DG/Fe_2O_3 electrode showed a high initial discharge capacity of 1,870.4 mAh·g^{-1} (Figure 2.10d). The discharge capacity decreased to 1,135.2 mAh·g^{-1} after the first five cycles and was well maintained in the following 130 cycles (Figure 2.10e). The 3DG/Fe_2O_3 electrode demonstrated good rate performance compared with pure Fe_2O_3 (Figure 2.10f). At a high current density of 5 A·g^{-1}, the 3DG/Fe_2O_3 electrode showed a discharge capacity of 523.5 mAh·g^{-1} with 98% capacity maintained even after 1,200 cycles (Figure 2.10g).

2.2.3 Carbon Cloth-Based Flexible Lithium-Ion Batteries

Compared with CNT and graphene, carbon cloth is also a good candidate for flexible electrodes for lithium-ion batteries, which shows high mechanical strength, electrical conductivity, flexibility, and relatively low cost. Based on carbon fiber cloth (CFC), highly crystalline VO_2 nanobelt arrays on CFC (CFC@VO_2, Figure 2.11a) were developed as a flexible cathode for lithium-ion batteries [52]. The CFC@VO_2 electrode showed well-distributed nanobelt arrays with average widths of 150–300 nm, lengths of 350–900 nm, and thicknesses of 50–100 nm (Figure 2.11b). The initial discharge capacity of CFC@VO_2 cathode was 145 mAh·g^{-1}, which was close to the theoretical capacity of 161 mAh·g^{-1} and was much higher than that of pure VO_2 (Figure 2.11c). The overpotential of the CFC@VO_2 cathode was 47 mV (Figure 2.11d), which was much lower than that of pure VO_2 (163 mV). The CFC@ VO_2 cathode demonstrated good rate performance compared with pure VO_2 (Figure 2.11e and f).

2.2.4 Other Flexible Lithium-Ion Batteries

Other different forms of conductive materials were also used in lithium-ion batteries. MXenes is a family of 2D transition metal carbides and nitrides with high electrical conductivity, mechanical property, and hydrophilicity, promising their application in energy storage devices. In 2020, through a simple vacuum filtration method, an MXene-bonded Si@C film as a flexible anode was made for lithium-ion batteries [53]. In the hybrid film electrode, the Si@C nanoparticles were embedded in the porous 3D structure of the MXene framework (Figure 2.12a). The MXene-bonded Si@C

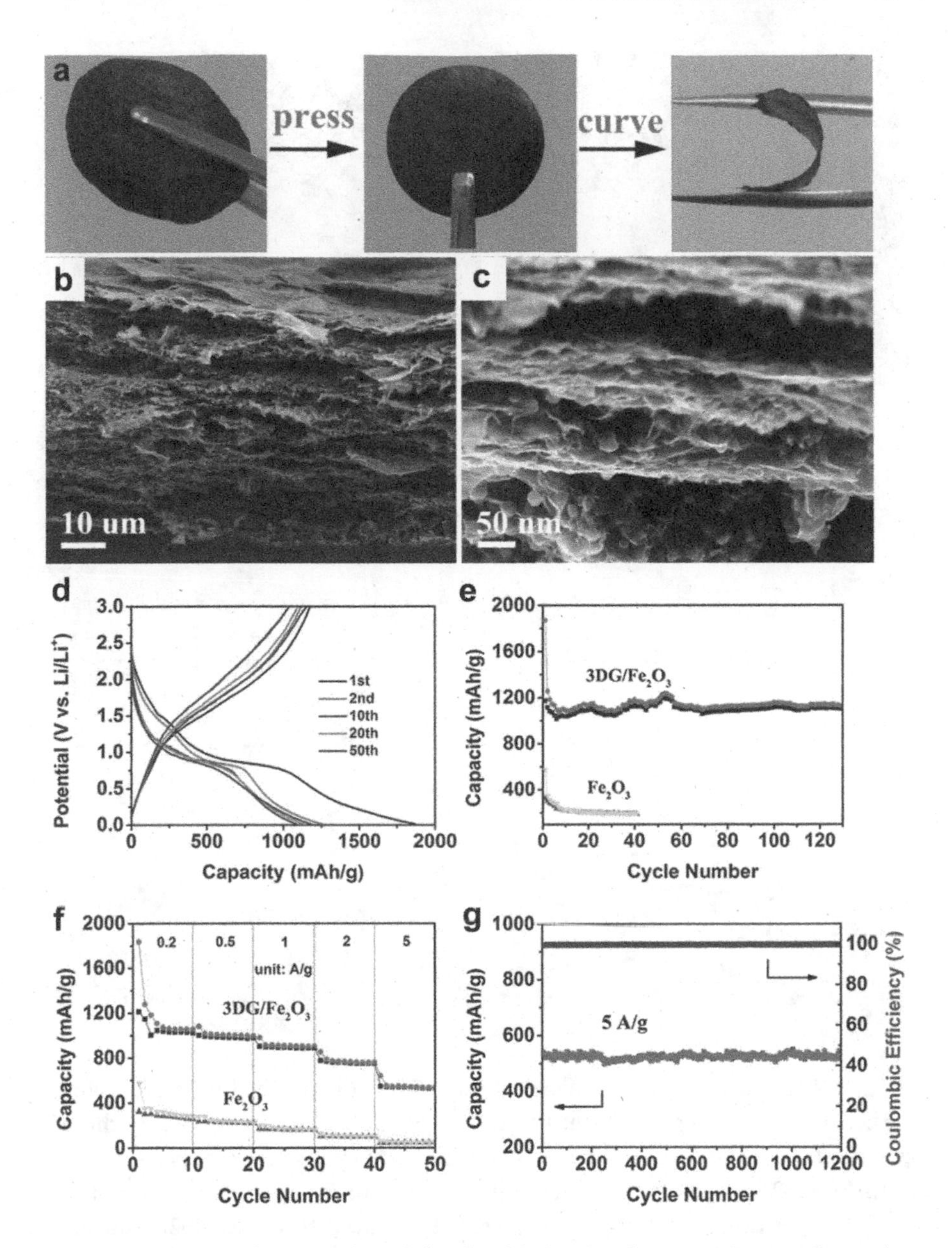

FIGURE 2.10 (a) Photographs of the flexible binder-free 3DG/Fe_2O_3 electrode. (b) SEM image of the pressed 3DG/Fe_2O_3 film. (c) SEM image of the internal microstructure of 3DG/Fe_2O_3 film. (d) Galvanostatic charge-discharge curves of the 3DG/Fe_2O_3 electrode at 0.2 $A \cdot g^{-1}$. (e and f) Cycling performance and rate performance of the 3DG/Fe_2O_3 and Fe_2O_3 electrodes at 0.2 $A \cdot g^{-1}$. (g) Cycling performance of the 3DG/Fe_2O_3 electrode at 5 $A \cdot g^{-1}$. (Reproduced from Ref. [51] with permission of the American Chemical Society.)

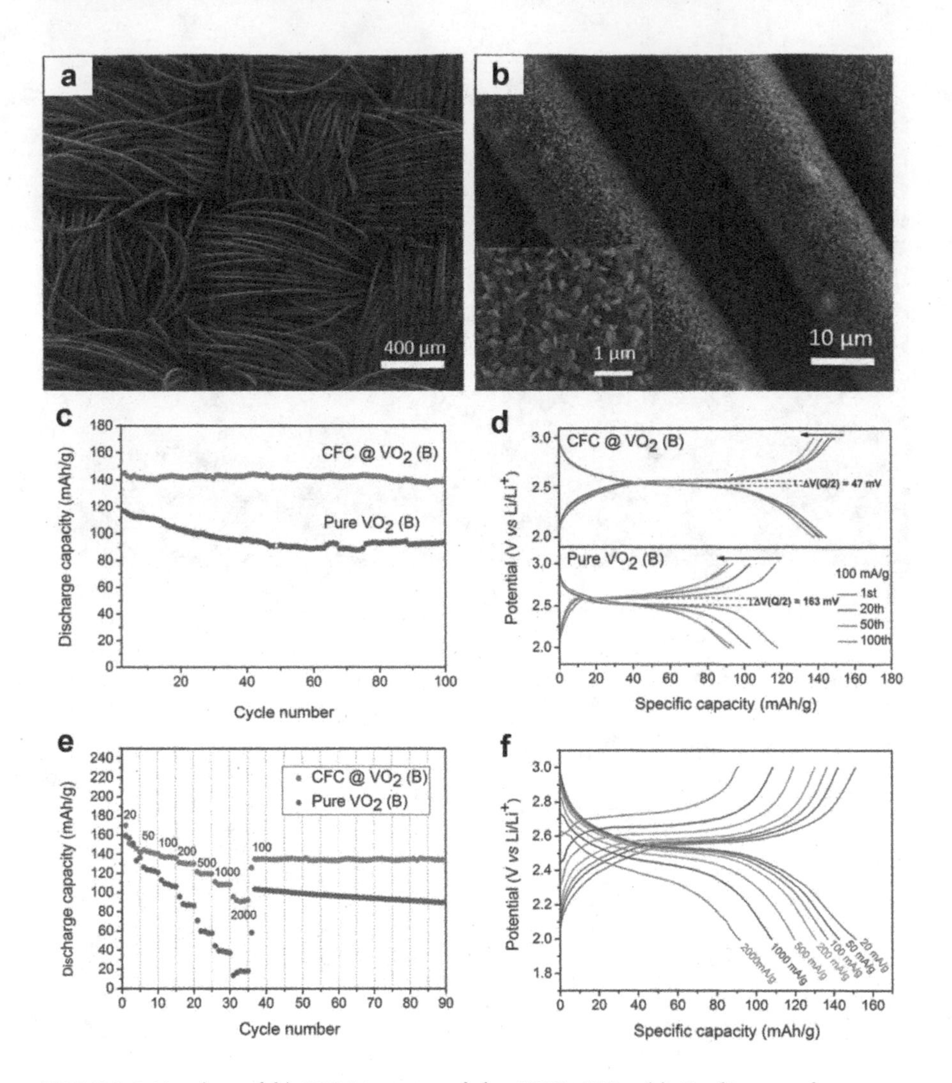

FIGURE 2.11 (a and b) SEM images of the CFC@VO_2. (c) Cycling performance of the CFC@VO_2 and pure VO_2 at 100 mA·g^{-1}. (d) Galvanostatic charge-discharge curves of the CFC@VO_2 and pure VO_2 at different cycles. (e) Rate performance of the CFC@VO_2 and pure VO_2. (f) Galvanostatic charge-discharge curves of the CFC@VO_2 at different current densities. (Reproduced from Ref. [52] with permission of the Royal Society of Chemistry.)

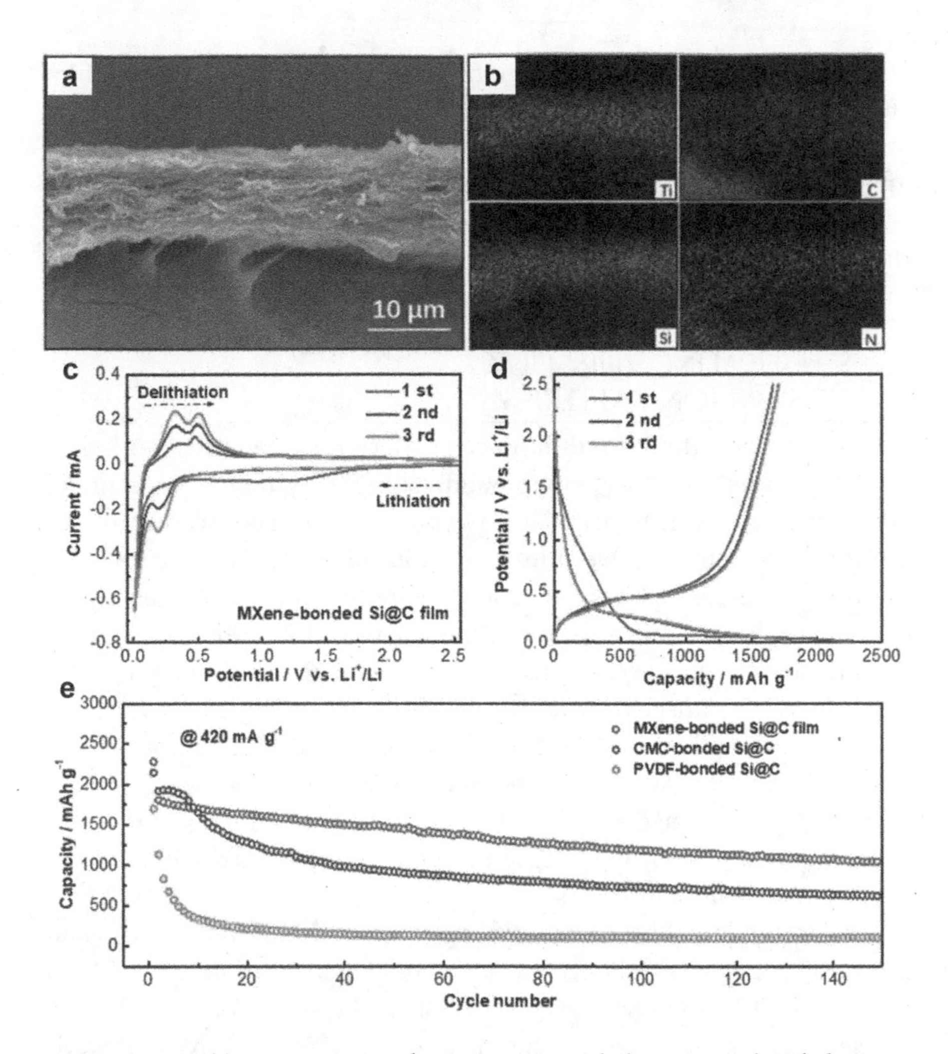

FIGURE 2.12 (a) Cross-sectional SEM image of the MXene-bonded Si@C film. (b) EDS mappings of the MXene-bonded Si@C film. (c) CV curves of the MXene-bonded Si@C film at 0.1 $mV{\cdot}s^{-1}$. (d) Galvanostatic charge-discharge curves of the MXene-bonded Si@C film at 420 $mA{\cdot}g^{-1}$. (e) Cycling performance of the MXene-bonded Si@C film at 420 $mA{\cdot}g^{-1}$. (Reproduced from Ref. [53] with permission of Wiley.)

electrode showed uniform distribution of Si@C nanoparticles in the film (Figure 2.12b). The MXene-bonded Si@C film showed initial discharge and charge capacities of 2,276.3 and 1,660.6 $mAh{\cdot}g^{-1}$, respectively, with an initial coulombic efficiency of 73.0% (Figure 2.12c and d). Besides, the MXene-bonded Si@C electrode demonstrated good cycling performance than CMC- and PVDF-bonded Si@C electrode during 150 cycles at 420 $mA{\cdot}g^{-1}$ (Figure 2.12e).

2.3 SELF-HEALING THIN-FILM LITHIUM-ION BATTERIES

Although flexible thin-film lithium-ion batteries can generally work during deformations such as bending and rolling, they might break under other complex deformations such as twisting that often occurs during use. On the other hand, flexible thin-film lithium-ion batteries usually have low thicknesses for high flexibility, leading to the damage and breaking of batteries under use. The breaking will make the lithium-ion batteries fail to work or even cause serious safety problems such as the leakage of toxic electrolytes. Organisms can heal mechanical injuries and recover body functions through the self-healing of the wound [54]. Therefore, if the flexible lithium-ion batteries can be endowed with a self-healing function, they will repair damages and break during their use in wearable applications, which will vastly increase the adaptability and reliability of the thin-film lithium-ion batteries.

A new kind of self-healing thin-film aqueous lithium-ion batteries had been realized by using aligned CNT sheets loaded with $LiMn_2O_4$ (LMO) and $LiTi_2(PO_4)_3$ (LTP) nanoparticles on a self-healing polymer substrate as electrodes and aqueous lithium sulfate/sodium carboxymethylcellulose (Li_2SO_4/CMC) as gel electrolyte [55]. The used self-healing polymer substrate was free-standing and rich in multiple hydrogen bonds (Figure 2.13a). The LMO and LTP nanoparticles were wrapped into the aligned CNT films to obtain the self-healing electrodes with a sandwiched structure (Figure 2.13b). The self-healing thin-film aqueous lithium-ion battery was then fabricated by pairing the self-healing cathode and anode with the Li_2SO_4/CMC gel electrolyte between them (Figure 2.13c).

After cutting into two separated parts, the self-healing thin-film aqueous lithium-ion battery can be healed to recover the normal functionality by simply contacting the two breaking parts for seconds, which can rapidly and efficiently solve the breaking problem during use (Figure 2.13d).

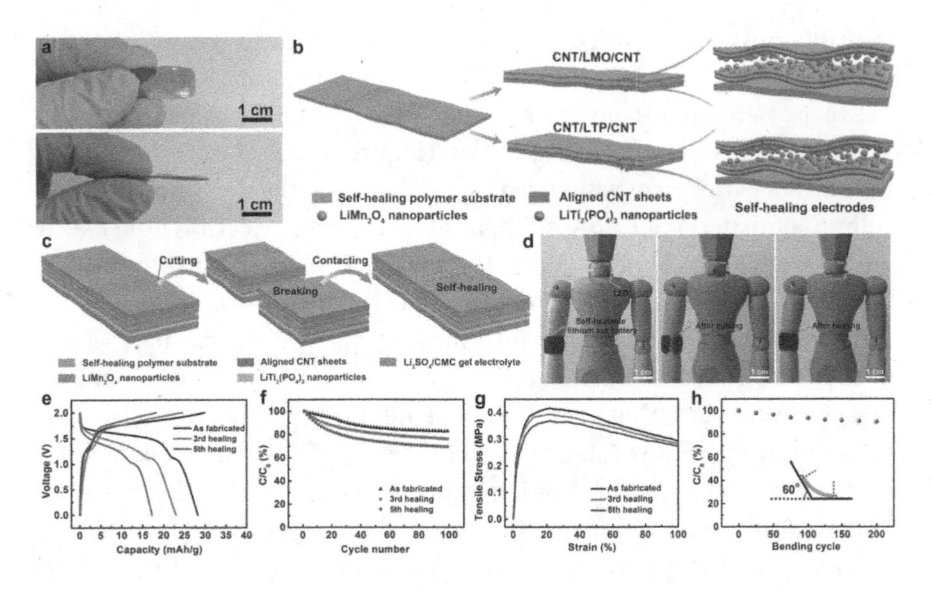

FIGURE 2.13 (a) Photographs of the self-healing polymer. (b) The fabrication process of the self-healing electrodes. (c) Schematic diagram of the self-healing process of thin-film aqueous lithium-ion battery. (d) Application demonstration of the self-healing aqueous lithium-ion battery worn around the elbow of a puppet. (e) Galvanostatic charge-discharge curves of the self-healing battery after different healing cycles (current density: 0.5 $A \cdot g^{-1}$). (f) Cyclic performance of self-healing battery after different healing cycles (current density: 0.5 $A \cdot g^{-1}$). (g) Stress-strain curves of the self-healing battery before and after healing. (h) The capacity retention of the self-healing aqueous battery under different bending cycles (bending angle: 60°). (Reproduced from Ref. [55] with permission of Wiley.)

Before cutting and after the fifth self-healing, the specific capacity of the thin-film battery decreased from 28.2 to 17.2 $mAh \cdot g^{-1}$ at the current density of 0.5 $A \cdot g^{-1}$ (Figure 2.13e). After the fifth healing, the specific capacity of the flexible battery was maintained by 69.3% after 100 cycles (Figure 2.13f). After the cutting-healing cycles, the mechanical strength of the self-healing thin-film battery was well maintained (Figure 2.13g). After 200 bending cycles at 60°, the specific capacity of the self-healing thin-film aqueous lithium-ion battery was maintained by 90.3% (Figure 2.13h).

An omni-healable thin-film aqueous lithium-ion battery was fabricated to self-repair the electrodes and electrolyte simultaneously after mechanical breakage [56]. The self-healable electrodes were fabricated by integrating the electroactive components (LiV_3O_8 and $LiMn_2O_4$) into polymer networks cross-linked by dynamic borate ester bonding (Figure 2.14a). The healed LiV_3O_8 gel electrode can be bent without breaking (Figure 2.14b), and there were almost no scars at the healed region, which proved the recovery of the electrode configuration (Figure 2.14c). The omni-healable lithium-ion battery was assembled by sandwiching the dopamine-grafted sodium alginate/Li_2SO_4 (SA-g-DA/Li_2SO_4) hydrogel electrolyte between LiV_3O_8 and $LiMn_2O_4$ electrodes (Figure 2.14d). After cutting/healing operations, the battery could power an LED (Figure 2.14e).

Before cutting and after the fifth healing, the specific capacity of the omni-healable thin-film aqueous lithium-ion battery decreased from 29.2 to 18.2 $mAh{\cdot}g^{-1}$ at the current density of 1.0 $A{\cdot}g^{-1}$ (Figure 2.14f). The battery's initial solution resistance and charge transfer resistance increased after the multiple cutting/healing process (Figure 2.14g). Besides, the omni-healable battery could be tailored without significant performance deterioration. The galvanostatic charge-discharge curves of the tailored battery were like that before tailoring and delivered a specific capacity of 22.7 $mAh{\cdot}g^{-1}$ at 1.0 $A{\cdot}g^{-1}$ (Figure 2.14h).

2.4 CHALLENGES AND PERSPECTIVE

With the rapid development of flexible and wearable electronics, flexible lithium-ion batteries are with excellent prospects and will promote the development of flexible electronics. The design of flexible structures can make the rigid lithium-ion batteries obtain a certain degree of flexibility, such as the thin-film structure. Flexible electrodes are critical for the realization of flexible batteries, and the design of flexible electrodes usually starts with current collectors. In this chapter, the advances in flexible thin-film lithium-ion batteries have been summarized from the current collectors based on CNT, graphene, carbon cloth, and MXene, and self-healing thin-film lithium-ion batteries are also highlighted for the next-generation direction.

Despite the significant advancement of flexible thin-film lithium-ion batteries, challenges still exist and hinder their applications. More efforts should be made for practical applications. First, the energy densities of the flexible thin-film lithium-ion batteries are still lower than those of conventional lithium-ion batteries, which need to be improved for better applications. Second, flexible thin-film lithium-ion batteries can generally work under general bending, and they might break under complex deformations like twisting that often occurs during use. Therefore, their flexibility

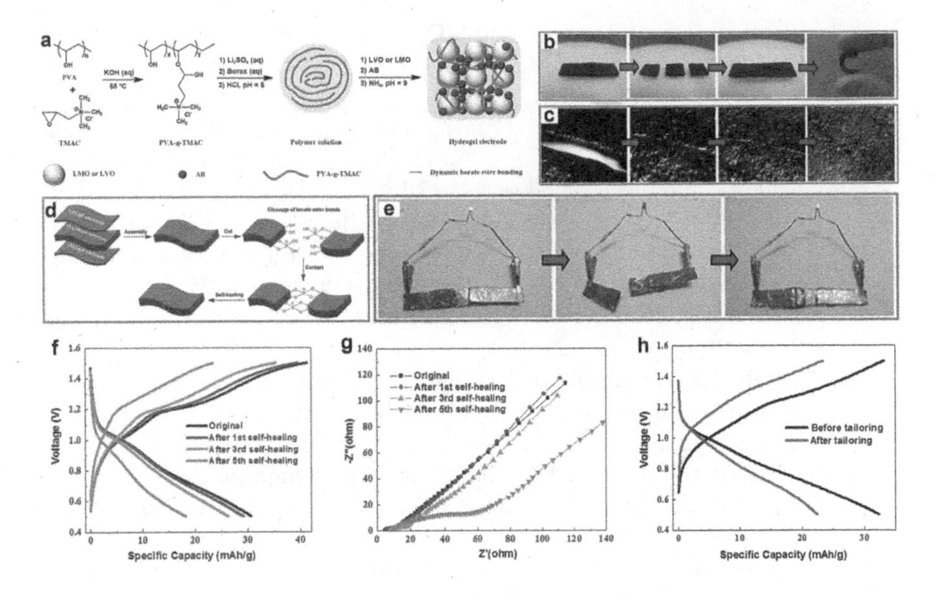

FIGURE 2.14 (a) Schematic diagram of the manufacturing process of the self-healable LiV_3O_8 and $LiMn_2O_4$ electrodes. (b) Optical micrographs of the self-healable electrode. (c) Micrographs of the self-healable electrode. (d) Schematic illustration of the assembly process and self-healing mechanism of the omni-healable thin-film aqueous lithium-ion battery. (e) Photographs of the omni-healable battery lighting a blue LED bulb before and after healing. Carbon paper was used to connect two batteries in series. (f) Galvanostatic charge-discharge curves of the healable battery (at 1 $A{\cdot}g^{-1}$) before and after different healing cycles. (g) Electrochemical impedance spectrum (EIS) spectra of the assembled battery before and after different times of healing. (h) Galvanostatic charge-discharge curves of the battery (at 1 $A{\cdot}g^{-1}$) before and after tailoring. (Reproduced from Ref. [56] with permission of Wiley.)

needs to be further improved to satisfy complex deformations. Third, as the production processes of flexible batteries are complicated, the costs of flexible thin-film batteries are higher than conventional lithium-ion batteries. It is thus necessary to develop large-scale and low-cost production methods for flexible thin-film lithium-ion batteries.

REFERENCES

1. Tarascon, J. M. Armand, M. 2001. Issues and challenges facing rechargeable lithium batteries. *Nature* 414: 359–367.
2. Armand, M. Tarascon, J. M. 2008. Building better batteries. *Nature* 451: 652–657.

3. Landi, B. J. Ganter, M. J. Cress, C. D. DiLeo, R. A. Raffaelle, R. P. 2009. Carbon nanotubes for lithium ion batteries. *Energy & Environmental Science* 2: 638–654.
4. Choi, J. W. Aurbach, D. 2016. Promise and reality of post-lithium-ion batteries with high energy densities. *Nature Reviews Materials* 1: 1–16.
5. Matsumura, Y. Wang, S. Mondori, J. 1995. Interactions between disordered carbon and lithium in lithium ion rechargeable batteries. *Carbon* 33: 1457–1462.
6. Xu, K. 2004. Non-aqueous liquid electrolytes for lithium-based rechargeable batteries. *Chemical Reviews* 104: 4303–4418.
7. Manthiram, A. 2020. A reflection on lithium-ion battery cathode chemistry. *Nature Communications* 11: 1–9.
8. Kraytsberg, A. Ein-Eli, Y. 2012. Higher, stronger, better…a review of 5 volt cathode materials for advanced lithium-ion batteries. *Advanced Energy Materials* 2: 922–939.
9. Wu, F. Maier, J. Yu, Y. 2020. Guidelines and trends for next-generation rechargeable lithium and lithium-ion batteries. *Chemical Society Reviews* 49: 1569–1614.
10. Babu, G. Kalaiselvi, N. Bhuvaneswari, D. 2014. Synthesis and surface modification of $LiCo_{1/3}Ni_{1/3}Mn_{1/3}O_2$ for lithium battery applications. *Journal of Electronic Materials* 43: 1062–1070.
11. Whittingham, M. S. 2004. Lithium batteries and cathode materials. *Chemical Reviews* 104: 4271–4302.
12. Li, Z. Wang, L. Li, K. Xue, D. 2013. $LiMn_2O_4$ rods as cathode materials with high rate capability and good cycling performance in aqueous electrolyte. *Journal of Alloys and Compounds* 580: 592–597.
13. Pan, F. Wang, W.-L. Li, H. Xin, X. Q. Yan, W. Chen, D. 2011. Investigation on a core-shell nano-structural $LiFePO_4$/C and its interfacial CO interaction. *Electrochimica Acta* 56: 6940–6944.
14. Jin, Y. Yang, C. Rui, X. Cheng, T. Chen, C. 2011. V_2O_3 modified $LiFePO_4$/C composite with improved electrochemical performance. *Journal of Power Sources* 196: 5623–5630.
15. Zhao, Y. Peng, L. Liu, B. Yu, G. 2014. Single-crystalline $LiFePO_4$ nanosheets for high-rate Li-ion batteries. *Nano Letters* 14: 2849–2853.
16. Zhang, W.-J. 2011. A review of the electrochemical performance of alloy anodes for lithium-ion batteries. *Journal of Power Sources* 196: 13–24.
17. Dey, A. 1971. Electrochemical alloying of lithium in organic electrolytes. *Journal of The Electrochemical Society* 118: 1547.
18. Etacheri, V. Marom, R. Elazari, R. Salitra, G. Aurbach, D. 2011. Challenges in the development of advanced Li-ion batteries: a review. *Energy & Environmental Science* 4: 3243–3262.
19. Szczech, J. R. Jin, S. 2011. Nanostructured silicon for high capacity lithium battery anodes. *Energy & Environmental Science* 4: 56–72.
20. Aurbach, D. Talyosef, Y. Markovsky, B. Markevich, E. Zinigrad, E. Asraf, L. Gnanaraj, J. S. Kim, H.-J. 2004. Design of electrolyte solutions for Li and Li-ion batteries: a review. *Electrochimica Acta* 50: 247–254.

21. Hassoun, J. Scrosati, B. 2015. Advances in anode and electrolyte materials for the progress of lithium-ion and beyond lithium-ion batteries. *Journal of the Electrochemical Society* 162: A2582.
22. Zheng, F. Kotobuki, M. Song, S. Lai, M. O. Lu, L. 2018. Review on solid electrolytes for all-solid-state lithium-ion batteries. *Journal of Power Sources* 389: 198–213.
23. Agapov, A. L. Sokolov, A. P. 2011. Decoupling ionic conductivity from structural relaxation: a way to solid polymer electrolytes? *Macromolecules* 44: 4410–4414.
24. Yao, X. Huang, B. Yin, J. Peng, G. Huang, Z. Gao, C. Liu, D. Xu, X. 2015. All-solid-state lithium batteries with inorganic solid electrolytes: review of fundamental science. *Chinese Physics B* 25: 018802.
25. Song, J. Wang, Y. Wan, C. C. 1999. Review of gel-type polymer electrolytes for lithium-ion batteries. *Journal of Power Sources* 77: 183–197.
26. Stephan, A. M. 2006. Review on gel polymer electrolytes for lithium batteries. *European Polymer Journal* 42: 21–42.
27. Cheng, X. Pan, J. Zhao, Y. Liao, M. Peng, H. 2018. Gel polymer electrolytes for electrochemical energy storage. *Advanced Energy Materials* 8: 1702184.
28. Zeng, W. Shu, L. Li, Q. Chen, S. Wang, F. Tao, X. M. 2014. Fiber-based wearable electronics: a review of materials, fabrication, devices, and applications. *Advanced Materials* 26: 5310–5336.
29. Zhong, J. Zhang, Y. Zhong, Q. Hu, Q. Hu, B. Wang, Z. L. Zhou, J. 2014. Fiber-based generator for wearable electronics and mobile medication. *ACS Nano* 8: 6273–6280.
30. Lee, J. Kwon, H. Seo, J. Shin, S. Koo, J. H. Pang, C. Son, S. Kim, J. H. Jang, Y. H. Kim, D. E. 2015. Conductive fiber-based ultrasensitive textile pressure sensor for wearable electronics. *Advanced Materials* 27: 2433–2439.
31. Zeng, L. Qiu, L. Cheng, H.-M. 2019. Towards the practical use of flexible lithium ion batteries. *Energy Storage Materials* 23: 434–438.
32. Fang, Z. Wang, J. Wu, H. Li, Q. Fan, S. Wang, J. 2020. Progress and challenges of flexible lithium ion batteries. *Journal of Power Sources* 454: 227932.
33. Hu, L. Hecht, D. S. Gruner, G. 2010. Carbon nanotube thin films: fabrication, properties, and applications. *Chemical Reviews* 110: 5790–5844.
34. He, Y. Chen, W. Gao, C. Zhou, J. Li, X. Xie, E. 2013. An overview of carbon materials for flexible electrochemical capacitors. *Nanoscale* 5: 8799–8820.
35. Zhou, G. Li, F. Cheng, H.-M. 2014. Progress in flexible lithium batteries and future prospects. *Energy & Environmental Science* 7: 1307–1338.
36. Wen, L. Li, F. Cheng, H. M. 2016. Carbon nanotubes and graphene for flexible electrochemical energy storage: from materials to devices. *Advanced Materials* 28: 4306–4337.
37. Zhang, X. Jiang, K. Feng, C. Liu, P. Zhang, L. Kong, J. Zhang, T. Li, Q. Fan, S. 2006. Spinning and processing continuous yarns from 4-inch wafer scale super-aligned carbon nanotube arrays. *Advanced Materials* 18: 1505–1510.
38. Chen, L. He, Y. Chai, S. Qiang, H. Chen, F. Fu, Q. 2013. Toward high performance graphene fibers. *Nanoscale* 5: 5809–5815.

39. Xu, Z. Liu, Z. Sun, H. Gao, C. 2013. Highly electrically conductive Ag-doped graphene fibers as stretchable conductors. *Advanced Materials* 25: 3249–3253.
40. Huang, S. Li, L. Yang, Z. Zhang, L. Saiyin, H. Chen, T. Peng, H. 2011. A new and general fabrication of an aligned carbon nanotube/polymer film for electrode applications. *Advanced Materials* 23: 4707–4710.
41. Liu, Z. Li, Z. Xu, Z. Xia, Z. Hu, X. Kou, L. Peng, L. Wei, Y. Gao, C. 2014. Wet-spun continuous graphene films. *Chemistry of Materials* 26: 6786–6795.
42. Xu, Y. Sheng, K. Li, C. Shi, G. 2010. Self-assembled graphene hydrogel via a one-step hydrothermal process. *ACS Nano* 4: 4324–4330.
43. Xie, X. Ye, M. Hu, L. Liu, N. McDonough, J. R. Chen, W. Alshareef, H. N. Criddle, C. S. Cui, Y. 2012. Carbon nanotube-coated macroporous sponge for microbial fuel cell electrodes. *Energy & Environmental Science* 5: 5265–5270.
44. Lota, G. Fic, K. Frackowiak, E. 2011. Carbon nanotubes and their composites in electrochemical applications. *Energy & Environmental Science* 4: 1592–1605.
45. Zhang, Q. Huang, J. Q. Qian, W. Z. Zhang, Y. Y. Wei, F. 2013. The road for nanomaterials industry: a review of carbon nanotube production, post-treatment, and bulk applications for composites and energy storage. *Small* 9: 1237–1265.
46. Park, S. Vosguerichian, M. Bao, Z. 2013. A review of fabrication and applications of carbon nanotube film-based flexible electronics. *Nanoscale* 5: 1727–1752.
47. Tao, T. Lu, S. Chen, Y. 2018. A review of advanced flexible lithium-ion batteries. *Advanced Materials Technologies* 3: 1700375.
48. Hu, L. Wu, H. La Mantia, F. Yang, Y. Cui, Y. 2010. Thin, flexible secondary Li-ion paper batteries. *ACS Nano* 4: 5843–5848.
49. Zhou, G. Wang, D.-W. Hou, P.-X. Li, W. Li, N. Liu, C. Li, F. Cheng, H.-M. 2012. A nanosized Fe_2O_3 decorated single-walled carbon nanotube membrane as a high-performance flexible anode for lithium ion batteries. *Journal of Materials Chemistry* 22: 17942–17946.
50. Abnavi, A. Faramarzi, M. S. Abdollahi, A. Ramzani, R. Ghasemi, S. Sanaee, Z. 2017. SnO_2@ a-Si core-shell nanowires on free-standing CNT paper as a thin and flexible Li-ion battery anode with high areal capacity. *Nanotechnology* 28: 255404.
51. Jiang, T. Bu, F. Feng, X. Shakir, I. Hao, G. Xu, Y. 2017. Porous Fe_2O_3 nanoframeworks encapsulated within three-dimensional graphene as high-performance flexible anode for lithium-ion battery. *ACS Nano* 11: 5140–5147.
52. Li, S. Liu, G. Liu, J. Lu, Y. Yang, Q. Yang, L.-Y. Yang, H.-R. Liu, S. Lei, M. Han, M. 2016. Carbon fiber cloth@VO_2 (B): excellent binder-free flexible electrodes with ultrahigh mass-loading. *Journal of Materials Chemistry A* 4: 6426–6432.
53. Zhang, P. Zhu, Q. Guan, Z. Zhao, Q. Sun, N. Xu, B. 2020. A flexible Si@C electrode with excellent stability employing an MXene as a multifunctional binder for lithium-ion batteries. *ChemSusChem* 13: 1621–1628.

54. Fuchs, Y. Brown, S. Gorenc, T. Rodriguez, J. Fuchs, E. Steller, H. 2013. Sept4/ARTS regulates stem cell apoptosis and skin regeneration. *Science* 341: 286–289.
55. Zhao, Y. Zhang, Y. Sun, H. Dong, X. Cao, J. Wang, L. Xu, Y. Ren, J. Hwang, Y. Son, I. H. 2016. A self-healing aqueous lithium-ion battery. *Angewandte Chemie International Edition* 55: 14384–14388.
56. Hao, Z. Tao, F. Wang, Z. Cui, X. Pan, Q. 2018. An omni-healable and tailorable aqueous lithium-ion battery. *ChemElectroChem* 5: 637–642.

CHAPTER 3

Flexible Fiber Lithium-Ion Batteries

3.1 OVERVIEW OF FIBER LITHIUM-ION BATTERIES

Fiber-based clothes have been used by human beings for thousands of years because they are soft, breathable, and durable, representing an ideal paradigm for developing wearable electronic devices [1]. The concept of a fiber lithium-ion battery can be traced as early as 2013 [2]. After that, a range of fiber batteries has been fabricated to show promise for wearable electronics applications. These batteries typically take the form of flexible fibers with diameters ranging from tens to hundreds of micrometers. Compared with the planar structure, the fiber shape offers many unique and promising advantages [3]. In addition to the typical bending deformation, the fiber batteries can accommodate complicated deformations, for example, twisting and stretching. These fiber batteries can be woven into fabrics through low-cost textile technology with porous structures to allow water vapor and air to transport through them, highly desired for wearable applications. Both fibers and textiles can effectively fit the curved surface of the human body. This chapter will take a general view of the emergence and evolution of fiber lithium-ion batteries.

3.2 FIBER ELECTRODE

Developing fiber electrodes is the key to the realization of fiber batteries. The fiber electrodes are generally constructed by incorporating

DOI: 10.1201/9781003273677-3

active materials into flexible and electrically conductive fiber substrates. Therefore, the fiber substrates are expected to exhibit high electrical conductivities to rapidly transfer electrons and a high specific surface area to load a high proportion of active materials. Effective methods should also be developed so that the electrochemically active materials can be stably anchored on the supporting substrates. Besides, the hybrid fibers should be highly flexible and stable to satisfy both the fabrication and the application.

3.2.1 Carbon Nanotube Fibers

Metal wires provide excellent electrical properties, but they are relatively heavy and rigid. In contrast, commercial polymer fibers are lightweight and flexible, but they are less conductive or even insulating. Recently, there are increasing interests in developing a new family of nanostructured fiber electrodes based on carbon nanotube (CNT) fibers [4]. CNTs, as previously discussed, show promising properties in electrical conductivity (10^5 $S{\cdot}cm^{-1}$), Young's modulus (0.9 TPa), and tensile strength (150 GPa) [5]. To translate these excellent properties of individual CNTs to the macroscopic scale, it is necessary to assemble CNTs with ordered structures, e.g., aligned CNT fiber.

3.2.1.1 Fabrication

A variety of methods have been developed to prepare macroscopic CNT fibers. The first widely used method is wet spinning [6]. CNTs were dispersed in an aqueous solution, then extruded out of a spinneret, and finally coagulated into a solid fiber by extracting the dispersant [7]. However, the intrinsic chemical inertness and interactions among CNTs limit their solubility in aqueous, organic, or acid media [8]. Therefore, surfactants or chemical treatments are necessary to increase compatibility with solvents. It seems unfeasible to prepare pure CNT fibers *via* this wet spinning technique as it contains impurities like surfactants and polymers, which are detrimental to the performances of CNTs, especially electrical conductivity. Besides, the wet spinning process typically produces a randomly distributed structure of CNTs, which produces low mechanical strength and electrical conductivity. As a result, increasing interests are attracted to prepare highly aligned CNT fibers by a drying spinning process [9–11].

Dry spinning produces aligned CNT fibers that exhibit advantages in preparation and morphology control (Figure 3.1). The CNT fiber is

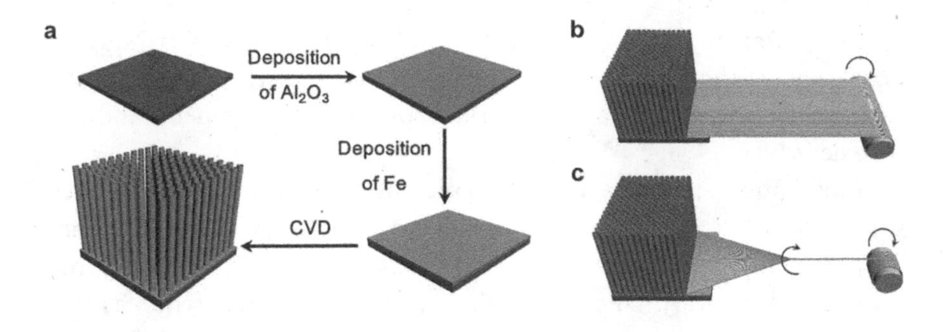

FIGURE 3.1 Schematic diagram of the fabrication of (a) spinnable carbon nanotube (CNT) array, (b) aligned CNT sheet, and (c) CNT fiber.

prepared from a unique CNT precursor—spinnable CNT array, where CNTs are highly aligned. Spinnable CNT arrays were synthesized by chemical vapor deposition (CVD). As shown in Figure 3.1a, the catalyst with a multilayered structure is deposited on the smooth surface of the substrate (e.g., Si and stainless steel). Typically, a buffer layer (Al_2O_3, ZnO, MgO, etc.) is first deposited to prevent the penetration of catalyst into the substrate and limit the mobility and Ostwald ripening of catalyst particles. Above the buffer layer, a catalyst film (Fe, Co, Ni, or their alloy) is deposited. The CVD process is carried out in a furnace flowed with a carbon source (methane, ethylene, acetylene, cyclohexane, etc.) and carrier gas (H_2/Ar). The growing temperature varies with different carbon sources.

When the temperature ramps up, catalyst film dewets into nanoparticles, and the nanoparticles are dispersed over the buffer layer. The diameter and density of catalyst particles depend on the thickness of catalyst film, ramping rate, carrier gas component, and compact buffer layer, and they further dictate the density and morphology of the resulting CNT array. During the later process, the carbon precursor is carried into the high-temperature zone of the furnace, where the carbon precursor is pyrolyzed and forms carbon clusters. The carbon clusters dissolve into the catalyst nanoparticles and form iron carbide in a liquid or liquid-like state. Once the catalyst is saturated, carbon atoms are prone to separate and nucleate, forming a carbon cap conforming to the shape of catalyst particles. Afterward, the carbon caps are lifted by the growing CNTs. The growth of CNT is accompanied by the formation of amorphous carbon that covers the catalyst particle and inhibits carbon dissolution. Consequentially, the growing process terminates. Since the morphology of the CNT array has a significant influence on its spinnability, the growth conditions of the spinnable CNT array are rigorous. The spinnable CNT array generally requires a clean surface, ordered arrangement, and suitable density.

The unique structure with suitable interaction and entanglement among neighboring CNT bundles makes this CNT array spinnable [12]. As shown in Figure 3.2a, the aligned CNT sheet is dry-spun from the CNT array [13]. Due to van der Waals interactions, the CNTs in the array are easy to aggregate into bundles, and single CNTs interpenetrates between two neighboring bundles. When a CNT bundle is being drawn out, CNTs penetrating the two neighboring bundles are divorced and gathered at the end of the bundles, which suffice for peeling off the neighboring CNT bundle. Subsequently, CNT bundles are peeled off and drawn out continuously. An aligned CNT fiber is produced by twisting the CNT sheet, where CNTs are packed at a helix angle with the fiber (Figure 3.2b and c). The twisted CNTs have a large entanglement which promises relatively high mechanical strength and porous structure.

3.2.1.2 Properties

The highly aligned structure of as-prepared CNT fiber offers high charge transport capability. The CNT fibers show conductivities of 10^4 S·cm^{-1} and display a three-dimensional hopping conduction mechanism, which is described by the following equation:

$$\sigma = \sigma_0 e^{-\frac{A}{T^{1/4}}}$$

where σ is the conductivity, σ_0 and A are constants, and T is the temperature. The electrical properties of CNT fiber can be further enhanced by posttreatment, such as thermal annealing and acid treatment [14].

The CNT fibers have a hierarchical structure, in which nanometer-sized building blocks are assembled into bundles, and the bundles are further assembled into macroscopically continuous fibers. The CNT fibers show tensile strengths of 10^2–10^3 MPa, which can be enhanced by optimizing the quality of the CNT array, the twisting process, and posttreatment. The CNT fibers have low bending stiffness (D) according to the equation,

$$D = \left(\pi \times d^3 \times E\right)/64$$

where d and E correspond to the fiber diameter and Young's modulus, respectively. In contrast to those of individual CNTs, the hierarchically aligned architecture of CNT fibers significantly reduces their Young's modulus. In addition, the tensile stresses in CNT fiber (void ratio of 0.35 or 0.44) are uniformly distributed along their length direction while severe stress concentration occurs in non-nanostructured fibers (void ratio of 0) under bending. Thus, the CNT fibers possess high flexibility and do not break under bending and knotting (Figure 3.2d).

Functional guests can be incorporated into the CNT fiber to extend its properties and applications [15]. For example, active materials can be stably anchored on the surfaces of aligned CNTs to favor the reversible intercalation and deintercalation of Li^+. The CNT fiber acts as a skeleton to support active materials and a current collector for charge transport. Besides, a variety of nanometer-sized and micrometer-sized hierarchical gaps are formed in the CNT fiber, favoring the diffusion of electrolytes to infiltrate through the gaps for rapid ion transport.

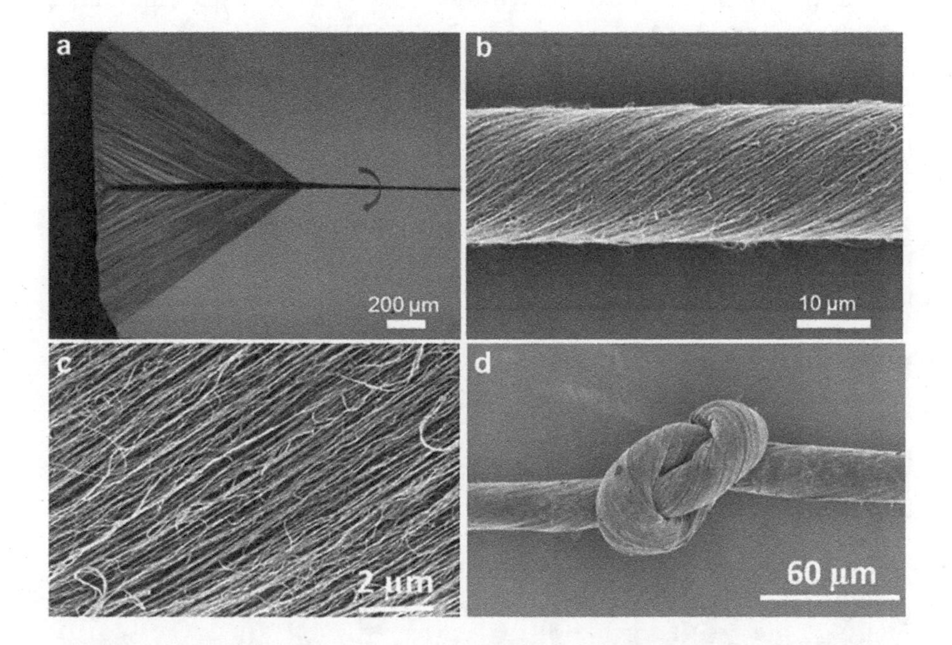

FIGURE 3.2 (a) SEM image of carbon nanotube (CNT) fiber dry-spun from CNT array. (b and c) SEM images of CNT fiber at low and high magnifications, respectively. (d) SEM image of a knotted CNT fiber. (Reproduced from Ref. [13] with permission of the American Chemical Society.)

3.2.2 Carbon Nanotube Hybrid Fiber Electrodes

Two effective methods have been developed to fabricate CNT hybrid fiber. The first method was *in situ* synthesis [16]. A variety of active materials are uniformly deposited on the surface of CNT after a chemical reaction, and the morphology of active materials can be well-tuned. The second method was co-spinning by a physical process. The suspension of active materials is first deposited onto the CNT sheets, followed by twisting into a hybrid fiber.

3.2.2.1 CNT/MnO_2 Fiber Electrode

The first CNT hybrid fiber electrode appeared in 2013, and it was synthesized by electrochemically depositing MnO_2 nanoparticles on the CNT fibers in an aqueous solution [2]. The weight percentages of the MnO_2 were controlled by the cycle number of depositions, ranging from 0.5% to 8.6%. MnO_2 nanoparticles were uniformly strewn on the surface of the CNT fiber (Figure 3.3). The hybrid fiber was flexible, and no noticeable decrease in structural integrity and electrical resistance was observed after bending for 100 cycles, indicating a stable attachment of MnO_2 nanoparticles on CNT fiber.

The electrochemical performance of CNT/MnO_2 hybrid fiber was further investigated with lithium wire as the anode in the 1 M $LiPF_6$ electrolyte solution. The average discharge platform was 1.5 V, and a specific capacity of 218.32 $mAh \cdot g^{-1}$ was achieved at 5×10^{-4} mA based on the whole weight of the hybrid fiber electrode. This work was a preliminary attempt toward flexible fiber-shaped lithium-ion batteries, and the performance needs to be further improved.

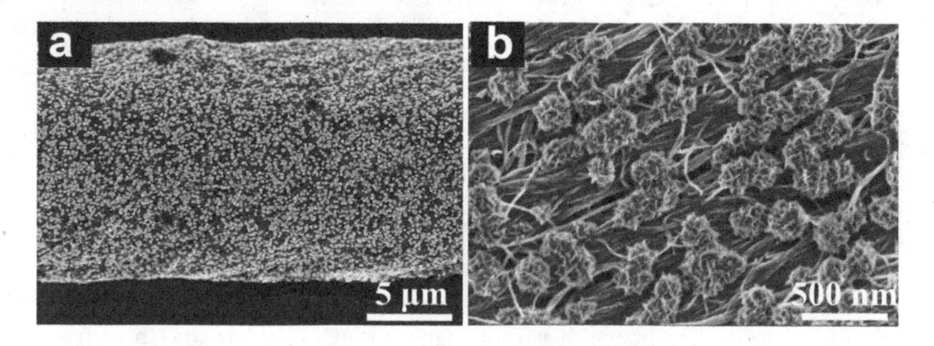

FIGURE 3.3 (a and b) SEM images of aligned carbon nanotube (CNT)/MnO_2 hybrid fibers with MnO_2 weight contents of 4.1% at low and high magnifications. (Reproduced from Ref. [2] with permission of Wiley-VCH.)

3.2.2.2 CNT/MoS_2 Fiber Electrode

As a typical two-dimensional layered transition metal sulfide, molybdenum disulfide (MoS_2) nanosheets show an energy storage capacity of 669 $mAh{\cdot}g^{-1}$, much higher than that of commercial graphite (372 $mAh{\cdot}g^{-1}$) [17]. However, MoS_2 usually suffers from structural deterioration due to significant volume changes during charge/discharge and low intrinsic electrical conductivity, leading to poor cyclability and rate capability [18].

Coupling MoS_2 with electrically conductive carbon is an effective method to overcome this obstacle. As shown in Figure 3.4, a CNT/MoS_2 hybrid fiber was fabricated [19]. MoS_2 nanosheets were hydrothermally synthesized on the surface of the CNT sheets, and then the hybrid sheets were twisted to form the CNT/MoS_2 hybrid fiber with a diameter of ~65 μm. The MoS_2 nanosheets were well distributed and wrapped on the surface of CNTs, and the CNTs remained highly aligned in the hybrid fiber (Figure 3.5). The high-magnification TEM image showed ultrathin MoS_2 nanosheets, providing a short length for the electrons to transport and an increased rate of lithium insertion and deinsertion. The structure of the hybrid fiber can be well maintained even after bending for 1,000 cycles, and the resistances were varied in less than 1% after bending for hundreds of cycles, which verified effective interaction between MoS_2 nanosheet and CNT.

The hybrid nanostructure efficiently combined high electrical conductivity in the CNT and high energy storage capacity in MoS_2. During the first discharge, the MoS_2 electrochemically dissociates irreversibly into Li_2S and Mo in a two-step process based on the following reactions:

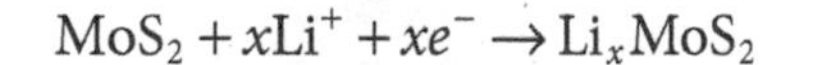

$$MoS_2 + xLi^+ + xe^- \rightarrow Li_xMoS_2$$

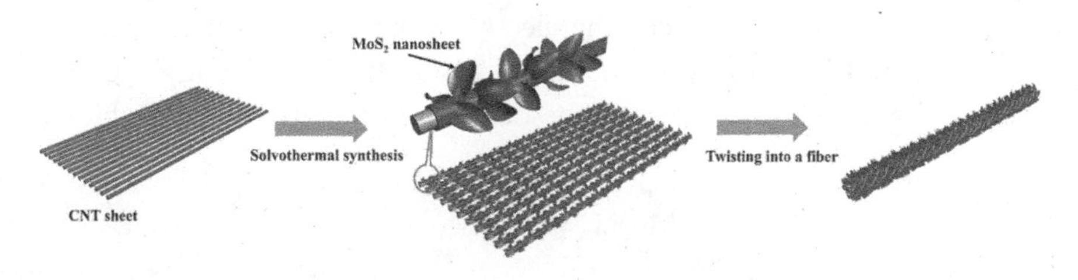

FIGURE 3.4 Schematic illustration of the synthesis of the carbon nanotube (CNT)/MoS_2 hybrid fiber. (Reproduced from Ref. [19] with permission of the Royal Society of Chemistry.)

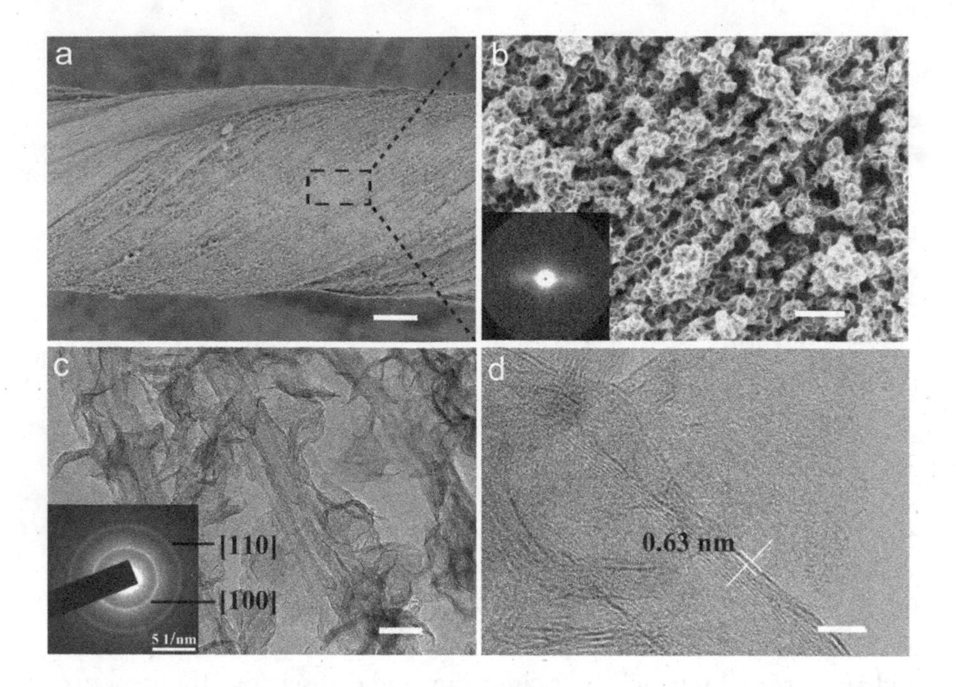

FIGURE 3.5 Morphology and structure of carbon nanotube (CNT)/MoS_2 hybrid fiber. (a and b) SEM images of CNT/MoS_2 hybrid fiber with increasing magnifications. The inserted image at (b) is the corresponding small-angle X-ray scattering pattern. (c) TEM image of CNTs coated with MoS_2 nanosheets with an inserted electron diffraction pattern. (d) Higher magnification of (c). Scale bars, 15 μm (a), 0.3 μm (b), 30 nm (c), and 5 nm (d). (Reproduced from Ref. [19] with permission of the Royal Society of Chemistry.)

$$Li_xMoS_2 + (4-x)Li^+ + (4-x)e^- \rightarrow Mo + 2Li_2S$$

During the first charge, Li_2S was oxidized into S and lithium ions. Therefore, after the first cycle, the electrode material was mainly composed of S and Mo instead of the initial MoS_2. In the following cycling, the conversion of S to Li_2S and the association of Li with Mo were able to react reversibly [20].

$$Li_2S + Mo/Li_x \rightleftharpoons S + Mo + Li_{x+2}$$

The specific capacity of the CNT/MoS_2 fiber electrode was maintained above 1,250 $mAh{\cdot}g^{-1}$ after 100 charge and discharge cycles. Effective interaction between MoS_2 and CNT makes the fiber electrode withstand

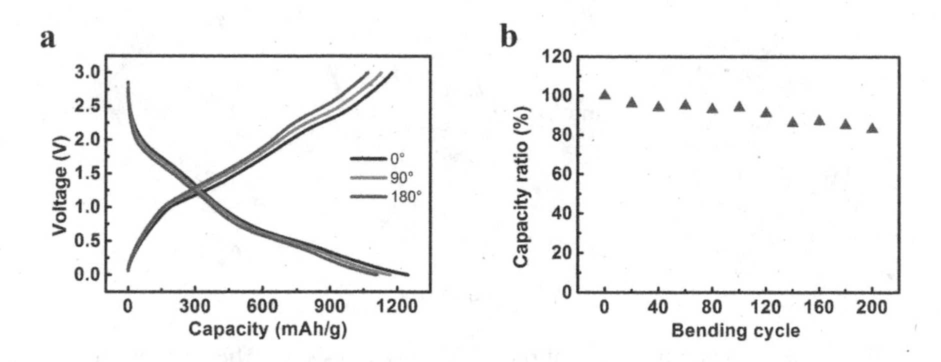

FIGURE 3.6 (a) Galvanostatic charge-discharge curves of carbon nanotube (CNT)/MoS_2 hybrid fiber before and after bending at 90° and 180°. (b) Dependence of specific capacity on bending cycle. C_0 and C correspond to the specific capacities before and after bending to 90° for different cycles. (Reproduced from Ref. [19] with permission of the Royal Society of Chemistry.)

bending deformations without sacrificing performance. As shown in Figure 3.6, the galvanostatic charge-discharge curves were perfectly overlapped at increasing bending angles, and the capacity was well maintained after bending for 200 cycles.

3.2.2.3 CNT/Si Fiber Electrode

Silicon, as an alloy-type anode material, has been widely studied due to several advantages [21,22]. First, silicon has a high gravimetric capacity of 4,200 $mAh{\cdot}g^{-1}$ upon full lithiation with the formation of $Li_{22}Si_5$. Second, the relatively low discharge potential plateau of 0.4 V *versus* Li^+/Li dramatically contributes to high working voltage paired with a cathode. One inevitable challenge with Si anode is their poor cycling stability due to the large volume changes up to 400%, resulting in electrode structure degradation, loss of contact points between the active material and conductive network, and unstable solid electrolyte interphase (SEI) formation.

Conductive carbon materials have been found vital for improving Si anodes, not only by the enhancement of the electrical conductivity of Si but also by buffering the volume change and stabilizing the SEI. A CNT/Si hybrid fiber was reported in 2014, and the fabrication was shown in Figure 3.7 [23]. Silicon was deposited onto the aligned CNT sheets through electron beam evaporation, and the amount of the silicon was controlled by the sputtering time. The hybrid sheets were further twisted into a hybrid CNT/Si hybrid fiber with a core-sheath structure.

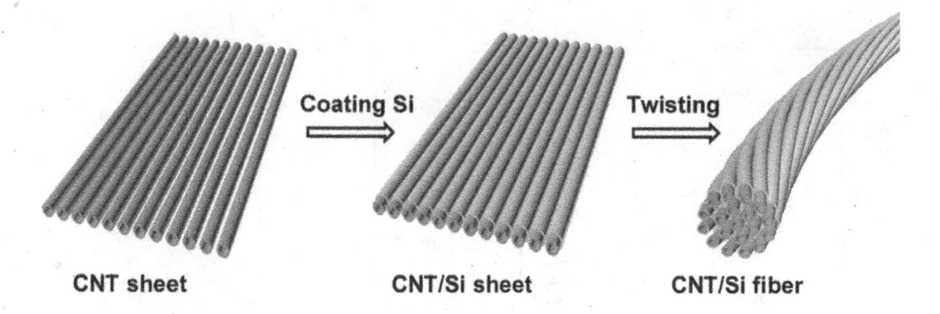

FIGURE 3.7 Schematic illustration of the synthesis of the carbon nanotube (CNT)/Si hybrid fiber. (Reproduced from Ref. [23] with permission of Wiley-VCH.)

The diameter of CNT fiber was increased from 30 to 60 μm after deposition of Si with a weight content of 38.1% (Figure 3.8). The aligned stricture was well maintained, and no apparent aggregates were observed in the hybrid fiber. The amorphous silicon with a thickness of ~20 nm was coated on the outer surfaces of the aligned CNTs. The core-sheath structure effectively exploited the high specific capacity of Si and the high electrical conductivity of CNT. Besides, the space among the aligned CNTs can effectively counterbalance the volume change. The hybrid fibers were also flexible and could be bent for over one hundred cycles without damage to structure integrity (Figure 3.9a and b).

The CNT/Si hybrid fiber was paired with a Li wire to characterize the electrochemical performance. The CNT/Si hybrid fiber showed a voltage plateau around 0.4 V vs. Li/Li^+. Compared with the CNT fiber, the CNT/Si hybrid fiber significantly increased the specific capacity. The capacities were increased from 82, 554, 1,090 to 1,670 $mAh{\cdot}g^{-1}$ with increasing weight percentages of Si from 0%, 18.7%, 26.7%, to 38.1%, respectively. The capacity was specified to the total amount of CNT and Si in the electrode. The CNT/Si hybrid fiber was flexible. After bending for 100 cycles, the specific capacity was maintained by 94% $mAh{\cdot}g^{-1}$ and survived more than 80% after 20 charge-discharge cycles (Figure 3.9).

However, the cycling performance of CNT/Si hybrid fiber was poor, and capacity was retained by 32% after 100 cycles. To further improve the cycling performance, a hybrid layered structure was developed [24]. Si was first deposited onto the CNT sheets by electron beam evaporation and then sandwiched between two bare CNT sheets. The hybrid sheets were finally scrolled into a hybrid fiber in a layered structure (Figure 3.10). The

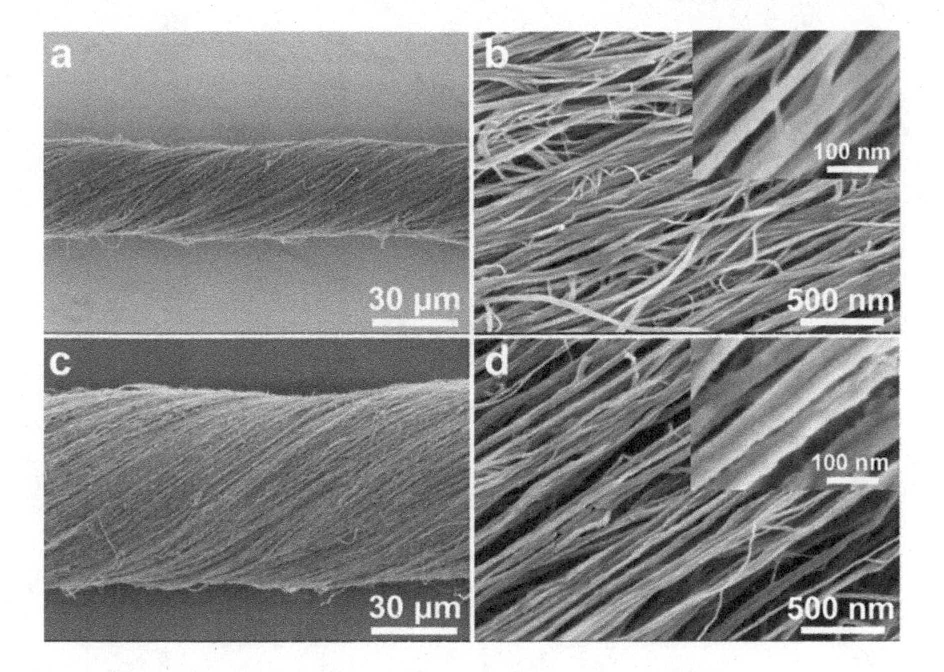

FIGURE 3.8 (a and b) SEM images of a bare aligned carbon nanotube (CNT) fiber at low and high magnifications, respectively. (c and d) SEM images of an aligned CNT/Si composite fiber with Si weight percentage of 38.1% at low and high magnifications, respectively. (Reproduced from Ref. [23] by permission of Wiley-VCH.)

diameter of the CNT/Si/CNT hybrid fiber was around 100 μm, and the content of Si was 62%. The initial delithiation capacity (charge capacity) was 2,240 mAh·g^{-1} at 0.4 C, which was retained by 88% after 100 cycles. The CNT skeleton provided effective conducting pathways making the fiber electrode adaptable for large currents. As a demonstration, the CNT/Si/CNT hybrid fiber delivered a delithiation capacity of 1,523 mAh·g^{-1} at a high current rate of 2 C (8,400 mA·g^{-1}), and more than 85% of the capacity survived after running for 400 cycles. The much-improved cycling performance was benefited from the hybrid layered structure where CNT sheets were conducive to immobilize the silicon layer and buffer the volume expansion.

3.2.2.4 CNT/$LiMn_2O_4$ and CNT/$Li_4Ti_5O_{12}$ Fiber Electrodes

Active materials have also been incorporated into CNT fibers *via* co-spinning by a physical process [25]. Compared with the *in situ*

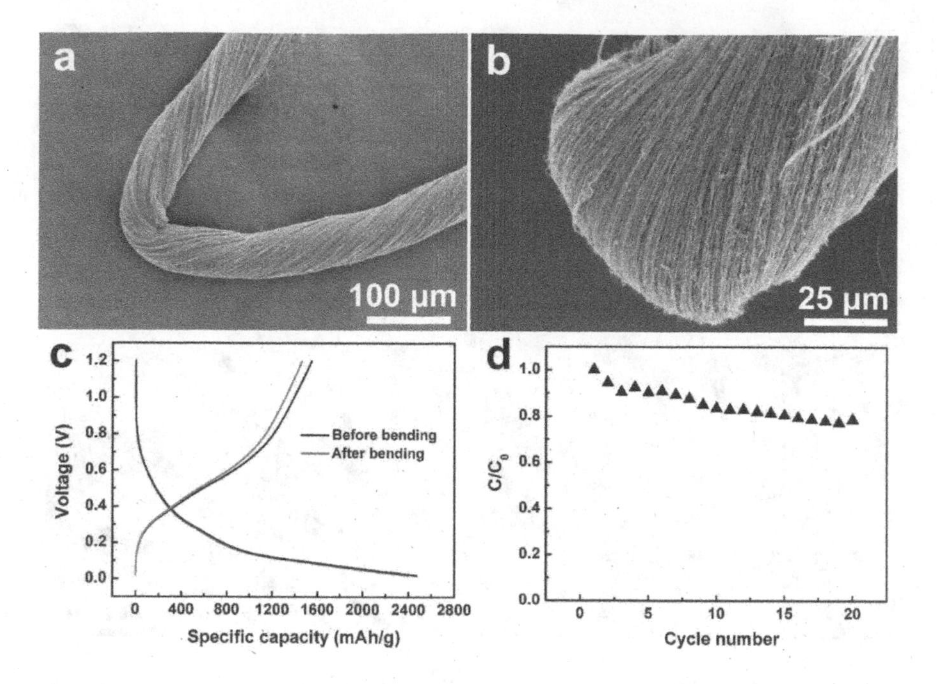

FIGURE 3.9 (a and b) SEM images of a bent aligned carbon nanotube (CNT)/Si hybrid fiber at low and high magnifications, respectively. (c) Charge and discharge curves of CNT/Si hybrid fiber before and after bending for 100 cycles at 2 $A{\cdot}g^{-1}$. (d) Dependence of specific capacity on cycle number for the hybrid fiber after bending for 100 cycles at 2 $A{\cdot}g^{-1}$. C_0 and C correspond to the specific capacities at the first and following cycles, respectively. (Reproduced from Ref. [23] with permission of Wiley-VCH.)

synthesis introduced before, the co-spinning method is a more general method, which almost applies to all kinds of nanoparticles. In a typical preparation, suspensions of active nanoparticles are deposited onto the aligned CNT sheets, followed by a scrolling process to produce the hybrid fibers (Figure 3.11a). The concentration of suspension controls the content of active materials, and this method can enable a high content of active nanoparticles.

Spinel $LiMn_2O_4$ is a commonly used economical cathode material capable of reversibly removing and restoring the utmost 95% of the theoretical capacity. $LiMn_2O_4$ is also the fastest cathode material due to its unique 3D structure, which allows multidirectional diffusion of Li^+ [26]. The SEM images of the CNT/$LiMn_2O_4$ hybrid fiber were displayed in Figure 3.11b and c. The CNT/$LiMn_2O_4$ hybrid fiber exhibited a uniform diameter of

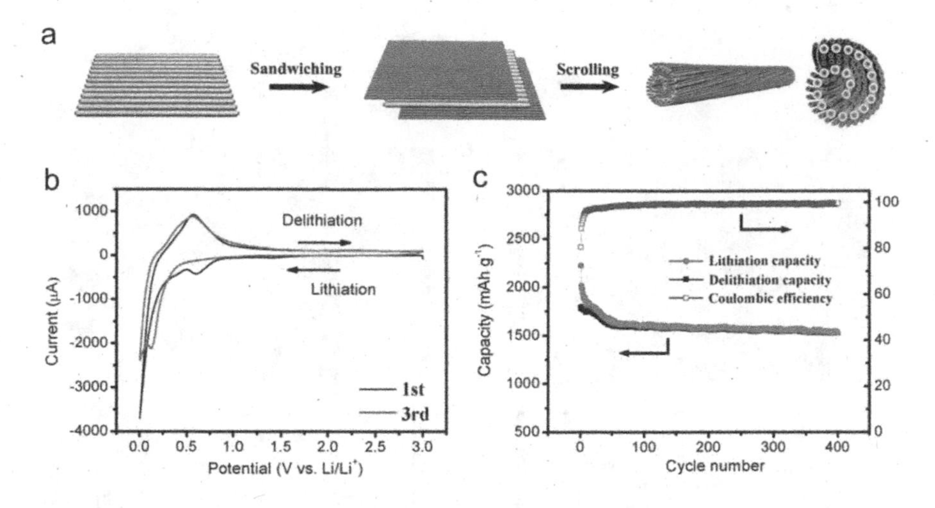

FIGURE 3.10 (a) Schematic illustration of the synthesis of the carbon nanotube (CNT)/Si/CNT hybrid fiber. (b) Cyclic voltammograms of CNT/Si/CNT hybrid fiber at a rate of 0.1 mV·s^{-1}. (c) Cycling performance of the CNT/Si/CNT hybrid fiber at 2 C. (Reproduced from Ref. [24] by permission of the American Chemical Society.)

~100 μm. The $LiMn_2O_4$ nanoparticles with an average diameter of 400 nm were uniformly dispersed and incorporated within the aligned CNT fiber. The weight percentage of $LiMn_2O_4$ was 75% in the hybrid fiber. The CNTs remained aligned after the co-spinning, which is advantageous to extend the excellent mechanical and electronic properties of individual CNTs to the macroscopic scale. The aligned and continuous CNTs functioned as effective pathways for charge transport. The hybrid fibers were flexible and robust, and they can be deformed into various shapes without obvious damages in structure.

Figure 3.11d shows typical charge and discharge curves of the aligned CNT/$LiMn_2O_4$ hybrid fibers at 0.5 C. Two distinct potential plateaus were observed in both charge and discharge curves. Two characteristic plateaus at 4.1 and 3.9 V represented a two-step process for the discharge process, i.e., from MnO_2 to $Li_{0.5}Mn_2O_4$ to $LiMn_2O_4$. The specific capacity was 100 mAh·g^{-1} based on the total weight of fiber electrode at 0.5 C and was maintained by over 94% after 100 cycles at 1 C.

Active materials such as $Li_4Ti_5O_{12}$ have been incorporated into an aligned CNT fiber *via* this co-spinning process. Spinel $Li_4Ti_5O_{12}$ has been previously investigated as a promising candidate since the insertion of Li

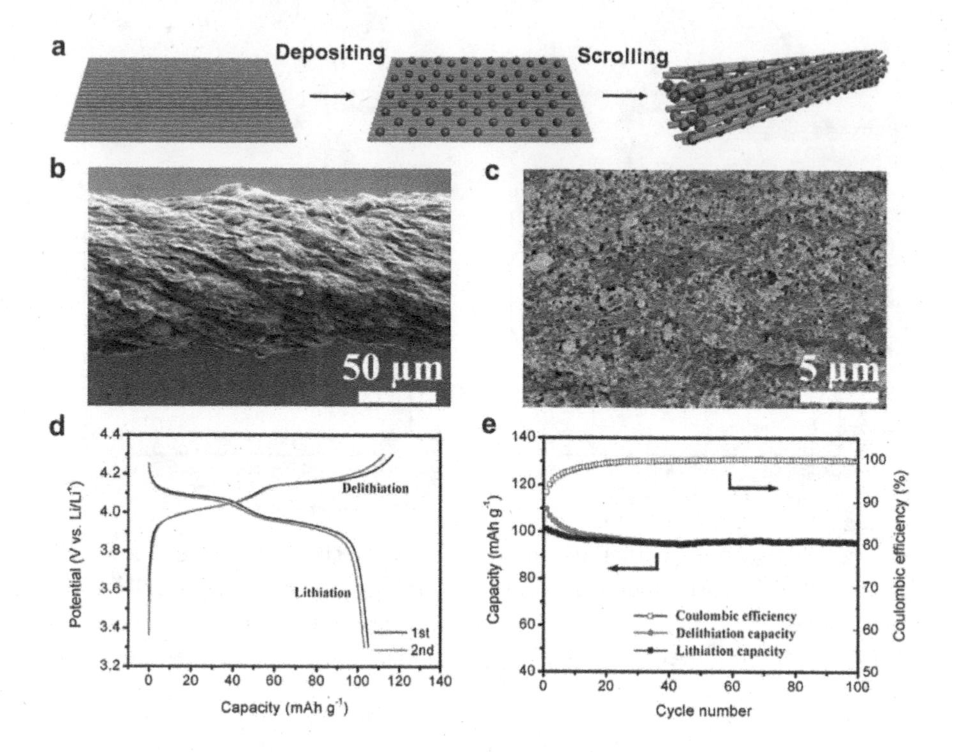

FIGURE 3.11 (a) Schematic illustration of the fabrication of the carbon nanotube (CNT)/$LiMn_2O_4$ hybrid fiber. (b and c) SEM images of CNT/$LiMn_2O_4$ hybrid fiber at low and high magnifications, respectively. (d and e) Charge-discharge profiles at 0.5 C. Cycling performance of CNT/$LiMn_2O_4$ hybrid fiber at 1 C. (Reproduced from Ref. [24] with permission of the American Chemical Society.)

ions can take place at ~1.5 V (vs. Li/Li^+) with a slight safety hazard [27]. In addition, the spinel $Li_4Ti_5O_{12}$ exhibits meager volume changes during cycling with remarkable stability. The CNT/$Li_4Ti_5O_{12}$ fiber electrode displayed a specific capacity of 150 $mAh{\cdot}g^{-1}$ and a discharge plateau of 1.5 V. The hybrid fiber could retain capacities of over 80% after 200 cycles.

3.3 FIBER LITHIUM-ION BATTERIES

Fiber lithium-ion batteries are generally constructed by twisting two fiber positive and negative electrodes. For example, the aligned CNT/$Li_4Ti_5O_{12}$ and CNT/$LiMn_2O_4$ hybrid fibers were assembled with a belt separator between them to form the fiber lithium-ion battery [28]. The battery was sealed in a heat-shrinkable tube with a diameter of 1.2 mm, as demonstrated in Figure 3.12. Lithium ions were removed from the

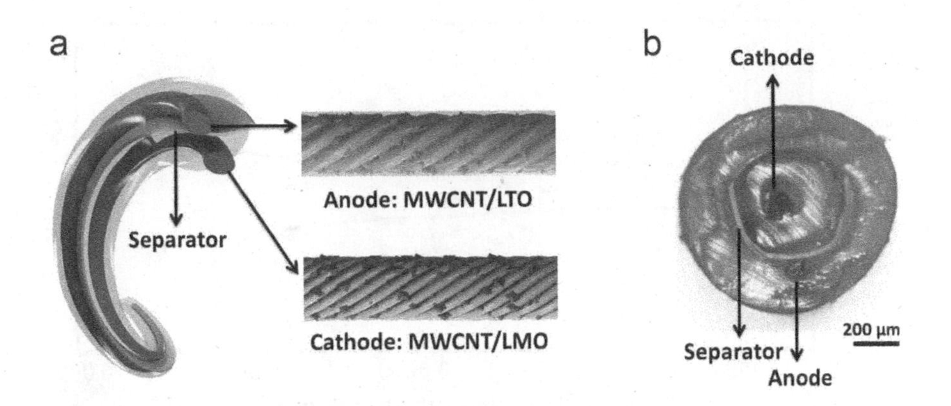

FIGURE 3.12 (a) Schematic diagram of the structure of the flexible fiber lithium-ion battery. The aligned carbon nanotube (CNT)/$Li_4Ti_5O_{12}$ and CNT/$LiMn_2O_4$ hybrid fibers were paired as the anode and cathode, respectively. (b) Cross-sectional micrograph of a full fiber battery. (Reproduced from Ref. [28] with permission of Wiley-VCH.)

$LiMn_2O_4$ lattice during the charge process, transferring into the electrolyte while intercalated into the $Li_4Ti_5O_{12}$ lattice at the anode. The process was reversed during discharge. Thus, the lithium ions shuttled between the two electrodes while the electrons, correspondingly, ran through the external circuit during the charge/discharge process.

$$\text{Reaction at the cathode: } LiMn_2O_4 \longleftrightarrow Li_{1-x}Mn_2O_4 + xLi^+ + xe^-$$

$$\text{Reaction at the anode: } Li\left[Li_{1/3}Ti_{5/3}\right]O_4 + xLi^+ \longleftrightarrow Li_{1+x}\left[Li_{1/3}Ti_{5/3}\right]O_4$$

The fiber battery achieved a high specific capacity of 0.0028 $mAh{\cdot}cm^{-1}$ (138 $mAh{\cdot}g^{-1}$) at 0.01 mA. The discharge plateau voltage was slightly decreased from 2.5 to 2.2 V with increasing currents from 0.01 to 0.05 mA. The fiber batteries delivered a discharge volumetric energy density of 17.7 $mWh{\cdot}cm^{-3}$ based on the total anode and cathode, higher than planar Li thin-film batteries (1–10 $mWh{\cdot}cm^{-3}$). The fiber battery also showed a stable galvanostatic charge-discharge cyclic performance, certified by 85% of capacity retention after 100 cycles at 0.05 mA (Figure 3.13).

The fiber battery was flexible and can be deformed into various shapes without physical damage and negligible performance losses (Figure 3.14). The voltage profiles were perfectly maintained even after 1,000 repeated bending tests. This fiber battery was woven into various flexible structures

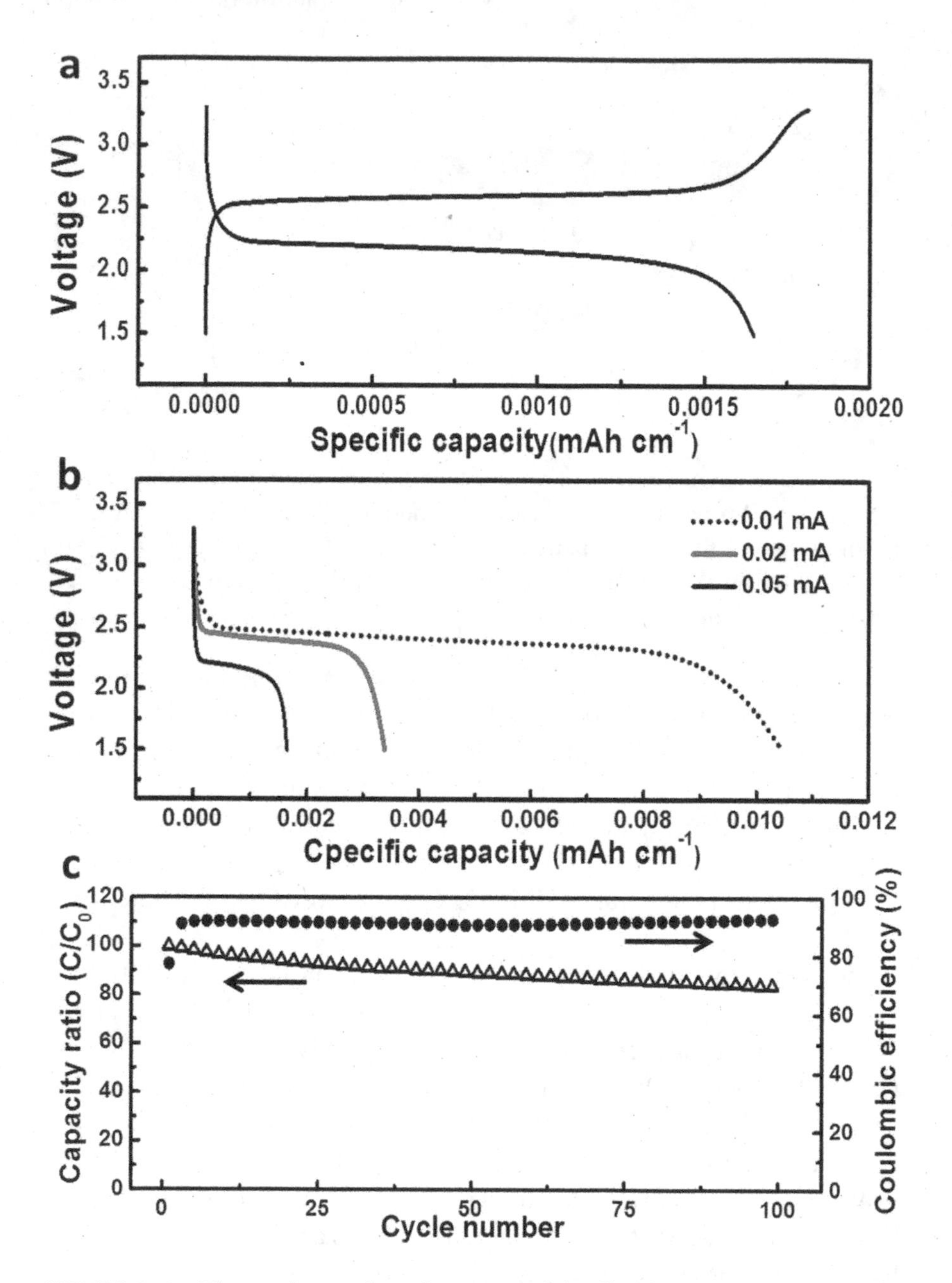

FIGURE 3.13 Electrochemical performance of the fiber battery with a length of 1 cm. (a) Galvanostatic charge and discharge curves at 0.05 mA. (b) Discharge profiles with increasing currents from 0.01, 0.02 to 0.05 mA. (c) Capacity retention and coulombic efficiency after 100 charge-discharge cycles at 0.05 mA. (Reproduced from Ref. [28] with permission of Wiley-VCH.)

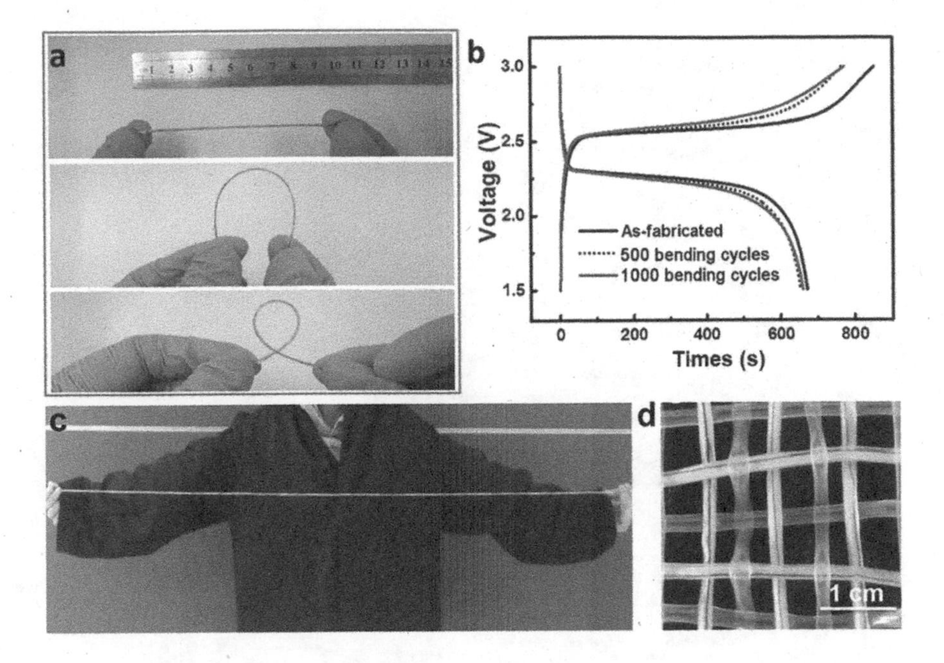

FIGURE 3.14 Flexibility of fiber lithium-ion batteries. (a) Photographs of a fiber battery being deformed into different formats. (b) Galvanostatic charge and discharge curves before and after bending for 500 and 1,000 cycles at 0.05 mA. (c) A fiber lithium-ion battery with a length of 200 cm. (d) Fiber batteries being woven into flexible textiles. (Reproduced from Ref. [28] with permission of Wiley-VCH.)

such as electronic textiles, which was expected to satisfy the ever-growing requirements for portable electronics.

Despite achievements made in developing fiber lithium-ion batteries in lab scale, their lengths were generally limited to several centimeters. Long fiber lithium-ion batteries were believed to have high internal resistance and not applicable as power sources. In 2021, Peng et al. revealed the hyperbolic cotangent function relationship of fiber lithium-ion batteries' internal resistance with their length based on the equivalent circuit model (Figure 3.15a). The measured resistance was in excellent accordance with the theoretical values (Figure 3.15b). Based on this discovery, the scale production of fiber lithium-ion batteries was enabled with industry-standard equipment and protocols, shown in Figure 3.15c. The as-prepared fiber lithium-ion batteries with lithium cobalt oxide as positive electrode and graphite as negative electrode were endowed with high cycling stability of 99.8% for 500 cycles with well-maintained charge-discharge voltage profiles

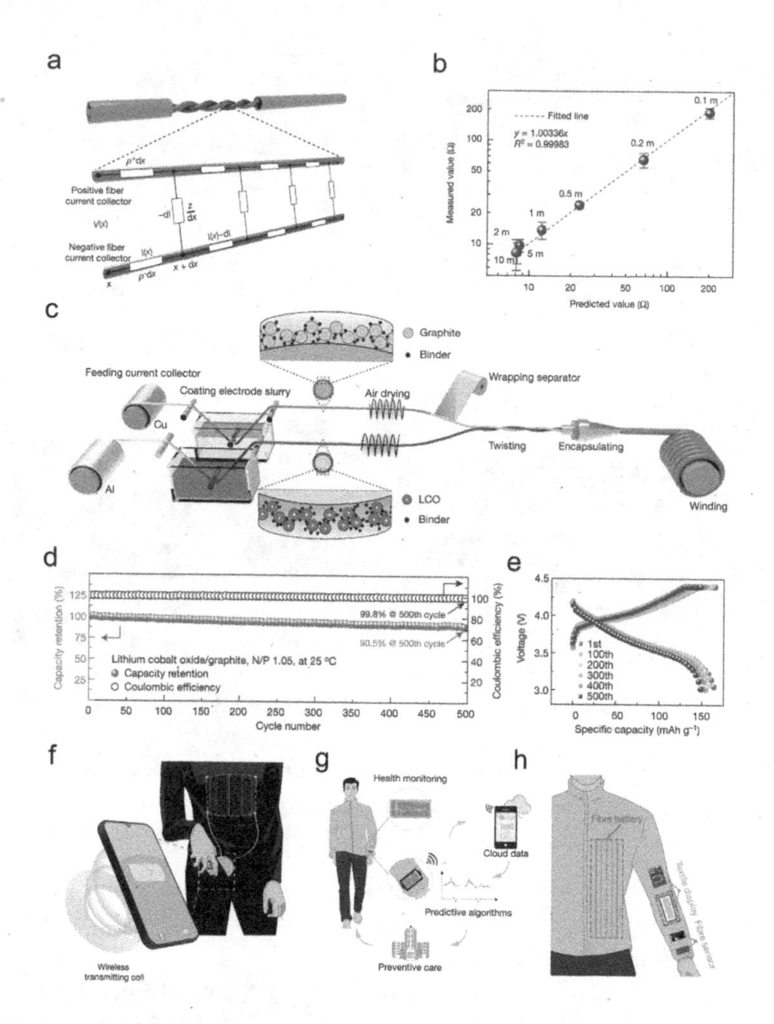

FIGURE 3.15 Scalable production of fiber lithium-ion batteries. (a) Equivalent circuit diagram of twisted fiber lithium-ion batteries, per unit length. (b) Measured internal resistances of fiber lithium-ion batteries correlate linearly (y=1.00336x; R^2=0.99983) with predicted values. (c) Schematic of the setup used to produce continuous fiber lithium-ion batteries. (d and e) Cycling performance, coulombic efficiency (c), and charge-discharge profiles (d) of fiber lithium-ion batteries show that capacity retention and coulombic efficiency remained high even after 500 cycles. (f) Illustration showing how fiber lithium-ion battery textile (gray dotted box) and a wireless transmitting coil (orange dotted box) integrated into everyday clothes can wirelessly charge a mobile phone. Inset: a mobile phone with a receiving coil obtains power from the transmitting coil. (g) Schematic illustrating an outdoor trial of a health management jacket integrated with fiber sensors for detecting ions in sweat and an electroluminescent textile display for displaying the data. (h) Block diagram shows how the FLIB powers the microcontroller and display driver. (Reproduced from Ref. [29] with permission of Nature Publishing Group.)

(Figure 3.15d and e). An outlook for practical applications of the flexible lithium-ion batteries were further explored. The fiber lithium-ion batteries could be woven into a textile and integrated into ordinary clothes, such as a jacket, to charge portal electronics, such as a smart phone, wirelessly (Figure 3.15f). The fiber lithium-ion batteries could also be integrated with health monitor systems with a textile display. The fiber lithium-ion batteries could power both the flexible sensors and textile display, which detected Na^+ and Ca^{2+} from sweat and exhibited the data in the textile display [29].

3.4 PERSPECTIVE

In summary, aligned CNT hybrid fibers were developed as effective electrodes for fabricating flexible fiber lithium-ion batteries, which opens up a new direction in the advancement of the next-generation electronics. Fiber lithium-ion batteries have made a series of significant advances in recent years. However, to truly realize the practical application, further research is needed in the following aspects. First, electrochemical performance needs to be further improved. Although fiber batteries have achieved high energy and power densities, the actual energy and power are low for these micro-batteries. Increasing the length of the battery is an effective way to improve energy and power. Second, currently reported fiber lithium-ion batteries are limited to lengths of centimeters. The subsequent research should focus on improving the electrical conductivity of the fiber electrode and optimizing its electrochemical performance to achieve a fiber lithium-ion battery with a length of up to meters. Besides, developing continuous preparation methods is essential to realize the large-scale production of fiber lithium-ion batteries.

REFERENCES

1. Zhang, Y. Zhao, Y. Ren, J. Weng, W. Peng, H. 2016. Advances in wearable fiber-shaped lithium-ion batteries. *Advanced Materials* 28: 4524–4531.
2. Ren, J. Li, L. Chen, C. Chen, X. Cai, Z. Qiu, L. Wang, Y. Zhu, X. Peng, H. 2013. Twisting carbon nanotube fibers for both wire-shaped micro-supercapacitor and micro-battery. *Advanced Materials* 25: 1155–1159.
3. Sun, H. Zhang, Y. Zhang, J. Sun, X. Peng, H. 2017. Energy harvesting and storage in 1D devices. *Nature Reviews Materials* 2: 1–12.
4. Zhang, Y. Jiao, Y. Liao, M. Wang, B. Peng, H. 2017. Carbon nanomaterials for flexible lithium ion batteries. *Carbon* 124: 79–88.
5. Yang, Z. Ren, J. Zhang, Z. Chen, X. Guan, G. Qin, L. Zhang, Y. Peng, H. 2015. Recent advancement of nanostructured carbon for energy applications. *Chemical Reviews* 115: 5159–5223.

6. Behabtu, N. Young, C. C. Tsentalovich, D. E. Kleinerman, O. Wang, X. Ma, A. W. K. Bengio, E. A. ter Waarbeek, R. F. de Jong, J. J. Hoogerwerf, R. E. Fairchild, S. B. Ferguson, J. B. Maruyama, B. Kono, J. Talmon, Y. Cohen, Y. Otto, M. J. Pasquali, M. 2013. Strong, light, multifunctional fibers of carbon nanotubes with ultrahigh conductivity. *Science* 339: 182–186.
7. Dalton, A. B. Collins, S. Munoz, E. Razal, J. M. Ebron, V. H. Ferraris, J. P. Coleman, J. N. Kim, B. G. Baughman, R. H. 2003. Super-tough carbon-nanotube fibres. *Nature* 423: 703.
8. Davis, V. A. Parra-Vasquez, A. N. G. Green, M. J. Rai, P. K. Behabtu, N. Prieto, V. Booker, R. D. Schmidt, J. Kesselman, E. Zhou, W. Fan, H. Adams, W. W. Hauge, R. H. Fischer, J. E. Cohen, Y. Talmon, Y. Smalley, R. E. Pasquali, M. 2009. True solutions of single-walled carbon nanotubes for assembly into macroscopic materials. *Nature Nanotechnology* 4: 830–834.
9. Jiang, K. Li, Q. Fan, S. 2002. Nanotechnology: spinning continuous carbon nanotube yarns. *Nature* 419: 801–801.
10. Li, Y. Kinloch, I. A. Windle, A. H. 2004. Direct spinning of carbon nanotube fibers from chemical vapor deposition synthesis. *Science* 304: 276–278.
11. Zhang, M. Atkinson, K. R. Baughman, R. H. 2004. Multifunctional carbon nanotube yarns by downsizing an ancient technology. *Science* 306: 1358–1361.
12. Kuznetsov, A. A. Fonseca, A. F. Baughman, R. H. Zakhidov, A. A. 2011. Structural model for dry-drawing of sheets and yarns from carbon nanotube forests. *ACS Nano* 5: 985–993.
13. Chen, T. Qiu, L. Cai, Z. Gong, F. Yang, Z. Wang, Z. Peng, H. 2012. Intertwined aligned carbon nanotube fiber based dye-sensitized solar cells. *Nano Letters* 12: 2568–2572.
14. Lu, W. Zu, M. Byun, J. H. Kim, B. S. Chou, T. W. 2012. State of the art of carbon nanotube fibers: opportunities and challenges. *Advanced Materials* 24: 1805–1833.
15. Lima, M. D. Fang, S. Lepro, X. Lewis, C. Ovalle-Robles, R. Carretero-Gonzalez, J. Castillo-Martinez, E. Kozlov, M. E. Oh, J. Y. Rawat, N. Haines, C. S. Haque, M. H. Aare, V. Stoughton, S. Zakhidov, A. A. Baughman, R. H. 2011. Biscrolling nanotube sheets and functional guests into yarns. *Science* 331: 51–55.
16. Zhang, Y. 2020. High-performance fiber-shaped lithium-ion batteries. *Pure and Applied Chemistry* 92: 767–772.
17. Zhou, J. Qin, J. Zhang, X. Shi, C. Liu, E. Li, J. Zhao, N. He, C. 2015. 2D space-confined synthesis of few-layer MoS_2 anchored on carbon nanosheet for lithium-ion battery anode. *ACS Nano* 9: 3837–3848.
18. Chen, C. Xie, X. Anasori, B. Sarycheva, A. Makaryan, T. Zhao, M. Urbankowski, P. Miao, L. Jiang, J. Gogotsi, Y. 2018. MoS_2-on-MXene heterostructures as highly reversible anode materials for lithium-ion batteries. *Angewandte Chemie International Edition* 57: 1846–1850.
19. Luo, Y. Zhang, Y. Zhao, Y. Fang, X. Ren, J. Weng, W. Jiang, Y. Sun, H. Wang, B. Cheng, X. Peng, H. 2015. Aligned carbon nanotube/molybdenum disulfide hybrids for effective fibrous supercapacitors and lithium ion batteries. *Journal of Materials Chemistry A* 3: 17553–17557.

20. Xiao, J. Wang, X. Yang, X. Xun, S. Liu, G. Koech, P. K. Liu, J. Lemmon, J. P. 2011. Electrochemically induced high capacity displacement reaction of PEO/MoS_2/graphene nanocomposites with lithium. *Advanced Functional Materials* 21: 2840–2846.
21. Feng, K. Li, M. Liu, W. Kashkooli, A. G. Xiao, X. Cai, M. Chen, Z. 2018. Silicon-based anodes for lithium-ion batteries: from fundamentals to practical applications. *Small* 14: 33.
22. Chae, S. Choi, S. H. Kim, N. Sung, J. Cho, J. 2020. Integration of graphite and silicon anodes for the commercialization of high-energy lithium-ion batteries. *Angewandte Chemie International Edition* 59: 110–135.
23. Lin, H. Weng, W. Ren, J. Qiu, L. Zhang, Z. Chen, P. Chen, X. Deng, J. Wang, Y. Peng, H. 2014. Twisted aligned carbon nanotube/silicon composite fiber anode for flexible wire-shaped lithium-ion battery. *Advanced Materials* 26: 1217–1222.
24. Weng, W. Sun, Q. Zhang, Y. Lin, H. Ren, J. Lu, X. Wang, M. Peng, H. 2014. Winding aligned carbon nanotube composite yarns into coaxial fiber full batteries with high performances. *Nano Letters* 14: 3432–3438.
25. Zhang, Y. Bai, W. Ren, J. Weng, W. Lin, H. Zhang, Z. Peng, H. 2014. Super-stretchy lithium-ion battery based on carbon nanotube fiber. *Journal of Materials Chemistry A* 2: 11054–11059.
26. Lee, S. Cho, Y. Song, H. Lee, K. T. Cho, J. 2012. Carbon-coated single-crystal $LiMn_2O_4$ nanoparticle clusters as cathode material for high-energy and high-power lithium-ion batteries. *Angewandte Chemie International Edition* 51: 8748–8752.
27. Yi, T. Yang, S. Xie, Y. 2015. Recent advances of $Li_4Ti_5O_{12}$ as a promising next generation anode material for high power lithium-ion batteries. *Journal of Materials Chemistry A* 3: 5750–5777.
28. Ren, J. Zhang, Y. Bai, W. Chen, X. Zhang, Z. Fang, X. Weng, W. Wang, Y. Peng, H. 2014. Elastic and wearable wire-shaped lithium-ion battery with high electrochemical performance. *Angewandte Chemie International Edition* 53: 7864–7869.
29. He, J. Lu, C. Jiang, H. Han, F. Shi, X. Wu, J. Wang, L. Chen, T. Wang, J. Zhang, Y. Yang, H. Zhang, G. Sun, X. Wang, B. Chen, P. Wang, Y. Xia, Y. Peng, H. 2021. Scalable production of high-performing woven lithium-ion fibre batteries. *Nature* 597: 57–63.

CHAPTER 4

Flexible Aqueous Lithium-Ion Batteries

4.1 OVERVIEW OF AQUEOUS LITHIUM-ION BATTERIES

Conventional lithium-ion batteries based on organic electrolytes have achieved sufficient energy and power densities for a variety of electronic devices [1,2]. However, the safety and nontoxicity of batteries have become a significant concern. Organic lithium-ion batteries suffer from serious safety risks, such as combustion and explosion at short circuit, as well as health concerns at electrolyte leakage. This situation is especially acute in flexible electronics as the chance of internal short circuit significantly increases under repeated deformations [3]. The manufacturing cost of organic lithium-ion batteries is driven up due to the high cost of raw materials and harsh production conditions. The assembly process of organic lithium-ion batteries requires a strict anhydrous environment due to the high reactivity of organic electrolytes with a trace amount of water in ambient air. Moreover, low ionic conductivity is generally observed for organic electrolytes [4].

By contrast, aqueous lithium-ion batteries which use water-based electrolytes eradicate these issues fundamentally. On the one hand, water-based electrolytes are generally composed of low-cost lithium salts (e.g., $LiNO_3$ or Li_2SO_4) dissolved in water with an ambient manufacturing environment. On the other hand, the aqueous electrolytes are endowed with conductivities two orders of magnitude higher than those of organic

DOI: 10.1201/9781003273677-4

electrolytes, contributing to high energy conversion efficiency and power densities [4,5].

Despite desirable advantages, the grid-scale applications of aqueous lithium-ion batteries are impeded by limited energy density and inferior cycle longevity, caused by the narrow voltage window of water (1.23 V) and side reactions of electrode and electrolyte. Hence, the principal component (i.e., electrodes and electrolytes) and energy storage mechanism of aqueous lithium-ion batteries have been globally investigated to overcome such problems and concluded as follows.

4.1.1 Electrodes

Figure 4.1 shows the stable electrochemical potential windows of water and the working voltages of electrode materials categorized as oxides, polyanions, and others under pH values ranging from 0 to 14. Due to their well-explored Li^+ storage capability and suitable working voltage (3–4 V vs. Li^+/Li), cathode materials primarily used in organic lithium-ion batteries are still promising for aqueous lithium-ion batteries while most conventional anode materials are no longer applicable due to their inappropriate working voltage (1–2 V vs. Li^+/Li) below H_2 evolution potential. Several

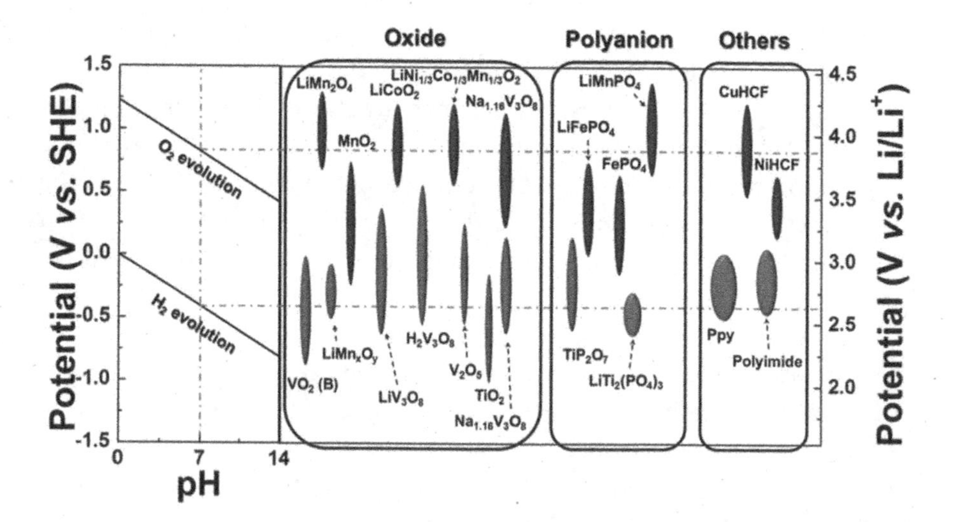

FIGURE 4.1 The stable electrochemical potential windows of water and three types of compounds used as active materials for aqueous lithium-ion batteries. (Reproduced from Ref. [4] with permission of the American Chemical Society.)

kinds of representative cathode materials and anode materials are introduced below.

4.1.1.1 Cathode Materials

Initially, $LiMn_2O_4$ (LMO) was used as the cathode with an average voltage of 1.5 V and an energy density of 75 $Wh·kg^{-1}$ [6]. Neutral lithium-salt electrolytes, such as LiCl and $LiNO_3$ aqueous electrolytes, are suitable for LMO [7,8]. Doping metal ions (Al, Cr) can improve the structural stability of LMO, thanks to the suppressed Jahn–Teller distortion [9,10]. The nanostructured LMO also exhibited high power and long cycle stability [11,12]. The nanostructures can reduce the diffusion length of Li^+ and increase the contact area of electrode and electrolyte, which can also compensate for the strain generated by Jahn–Teller deformation and inhibit the dissolution of Mn in the electrolyte.

The extraction and insertion of Li^+ into $LiCoO_2$ are more viable to be realized in $LiNO_3$ and Li_2SO_4 electrolytes [13,14]. The $LiCoO_2$ electrode exhibited a redox reaction at around ~0.9 V (vs. standard hydrogen electrode, SHE) and a capacity of 115 $mAh·g^{-1}$ in 5 M $LiNO_3$ electrolyte. Nano-$LiCoO_2$ can effectively improve the rate capability of aqueous lithium-ion batteries from a combination of size reduction and rapid ion transport in an aqueous electrolyte [15].

$LiFePO_4$ electrode delivered a redox reaction at around ~0.2 V (vs. standard calomel electrode, SCE) and a capacity of 130 $mAh·g^{-1}$ in 0.5 M Li_2SO_4 electrolyte. However, there was a rapid capacity fading of the $LiFePO_4$ electrode due to side reactions caused by O_2 and OH^-. After carbon coating, the $LiFePO_4$/C composite delivered improved cycle stability since the carbon protection could effectively block the attack of O_2 and OH^- [16].

4.1.1.2 Anode Materials

VO_2(B) with tunnel structure is conducive to the rapid intercalation and deintercalation of Li^+. The redox potential of VO_2(B) at −0.43 V vs. SHE makes it suitable for electrolytes with pH higher than 7.29 (hydrogen evolution potential was −0.43 V vs. SHE at pH 7.29) [6]. However, in a strongly alkaline electrolyte (pH 11.3), the capacity decreased rapidly with the VO_2(B) matrix dissolution. The VO_2(B) electrode exhibited good capacity retention under moderate alkaline conditions (pH 8.2) [17]. VO_2(B) with higher capacity and better cycling stability can be obtained by morphology optimization or carbon coating [18,19].

Layered γ-LiV_3O_8 exhibited a redox potential of −0.1 V vs. SHE in 1 M Li_2SO_4 or LiCl electrolyte at a pH of 6.2 [20]. The deterioration of the crystal structure of the γ-LiV_3O_8 anode led to its poor capacity retention [21]. Nanostructured $Li_{1+X}V_3O_8$ could utilize 70% of its theoretical capacity [22]. Controlling the dissolution of γ-LiV_3O_8 by eliminating oxygen in electrolytes and using a high concentration of lithium salt (9 M $LiNO_3$) can significantly improve the cyclic stability of aqueous lithium-ion batteries [23].

V_2O_5 has a high theoretical capacity and high electronic conductivity based on the mixing valence of vanadium [24]. The redox potential of V_2O_5 was evident at about −0.3 V vs. SHE. However, the dissolution of vanadium ions with crystal structure changed and amorphization during the cycle led to poor capacity retention [25]. Surface coatings, such as polypyrrole and polyaniline, can inhibit the dissolution of vanadium ions, slow down the amorphization of V_2O_5, and enhance the cycle stability of the battery [25,26].

$LiTi_2(PO_4)_3$ (LTP) demonstrated redox reaction at −0.45 V vs. SHE and discharge capacity of about 115 $mAh \cdot g^{-1}$ [27]. However, the capacity decayed rapidly due to the unstable crystal structure. Similarly, after coating a carbon layer on LTP, its capacity retention increased to 80% after 200 cycles [28]. Besides, it is essential to maintain the neutral pH of the electrolyte and eliminate water decomposition to improve the cycle stability [29].

Other anode materials for aqueous lithium-ion batteries include electroactive organic molecules and polymers storing Li^+ and electrons in their conjugated chemical bonds [30,31]. Polypyrrole anode could perform reversible "doping and dedoping" of Li^+ at −0.27 V vs. SCE [32]. Polyimide stored Li^+ through the charge distribution of conjugated carbonyl groups ($((C{=}O)_{2n})$) in aromatic molecules [33]. The polyimide anode showed a voltage level of −0.5 V vs. SCE and a capacity of 160 $mAh \cdot g^{-1}$ in 0.5 M $LiNO_3$ electrolyte. In addition, aqueous lithium-ion batteries with polyimide anodes exhibited good power capacity due to the absence of crystal distortion and strain caused by the diffusion of Li^+ through the host crystal structure.

4.1.2 Electrolytes

Electrolytes are crucial to improve the energy densities of aqueous lithium-ion batteries. Aqueous electrolytes usually consist of water as solvent and Li salts ($LiClO_4$, $LiNO_3$, Li_2SO_4, LiCl, $LiPF_6$, $LiBF_4$, and LiTFSI).

Although the pH of the electrolytes can adjust the hydrogen evolution reaction (HER)/oxygen evolution reaction (OER) potentials of water, the overall electrochemical stability windows of water are constant. In organic lithium-ion batteries, a solid electrolyte interface film is typically generated by the decomposition of organic electrolytes, providing an additional interface between the electrolytes and electrodes and expanding the available electrochemical stabilization windows of the electrolytes. However, a lack of solid electrolyte interface film in aqueous lithium-ion batteries due to the decomposition products of water (H_2, O_2, or OH^-) cannot form a deposition layer on the electrode surface.

Hence, "water-in-salt" electrolytes are proposed to broaden the voltage window of aqueous lithium-ion batteries effectively. In a water-in-salt electrolyte, salt outnumbers water by both weight and volume. The water-in-salt electrolyte prepared by lithium bis(trifluoromethane)-sulfonimide (LiTFSI) salt with high water solubility and high hydrolytic stability successfully expands electrolyte stability by limiting the number of free water molecules and forming a solid electrolyte interface [34]. For the anode, the reduction of TFSI resulted in a passivation process that intensified with salt concentration and reduced the plateau current. This passivation suppressed the H_2 evolution eventually and pushed its onset from 2.63 to 1.90 V vs. Li^+/Li. For cathode, O_2 evolution also was suppressed from 3.86 to 4.90 V vs. Li^+/Li with increasing salt concentrations due to the reduced water activity. As a result, a stable electrochemical window of ~3.0 V was achieved with 21 M LiTFSI.

However, 21 M LiTFSI is already near its saturation point at room temperature, in which there is no room for any solid electrolyte interface improvement by further increasing the salt concentration. In order to overcome the saturation limit of LiTFSI, the "water-in-bisalt" electrolytes (21 M LiTFSI-7 M lithium trifluoromethane sulfonate) [35], hydrate melt (19.4 M LiTFSI-8.3 M lithium bis(pentafluoroethanesulfonyl)-imide), [36] and monohydrate melt (22.2 M LiTFSI-33.3 M lithium (trifluoromethanesulfonyl) (pentafluoroethanesulfonyl) imide) [37] were developed to further broaden the electrochemical window. In addition, to address the high cost and toxicity of lithium fluoride compound, concentrated mixed cationic acetate water-in-salt electrolyte (32 M potassium acetate-8 M lithium acetate) [38] and "molecule crowded" aqueous electrolyte (poly(ethylene glycol)-2 M LiTFSI) [39] were proposed. Despite high concentrations, these electrolytes are still fluids, and there is a risk of leakage when used in flexible lithium-ion batteries. Therefore, hydrogel electrolyte is a good strategy for developing flexible aqueous lithium-ion batteries, e.g., water-in-ionomer electrolytes (50% lithiated polyacrylic acid-50% H_2O) [40] can be adopted to improve the voltage windows of aqueous lithium-ion batteries.

4.1.3 Working Mechanism

The aqueous lithium-ion batteries exhibited a similar working mechanism with organic lithium-ion batteries. Li^+ extracted from the anode during the discharging process migrated through the aqueous electrolyte and inserted into the cathode. At the same time, the electrons flowed from anode to cathode through the external circuit.

Different from organic lithium-ion batteries, there is water splitting in aqueous lithium-ion batteries, OER and HER. The potential difference between the two reactions is only 1.23 V. The regular operation of the aqueous lithium-ion batteries must ensure that the Li^+ extraction/insertion proceeds beyond water splitting. Otherwise, OER and HER will take place in priority, which not only lead to continuous consumption of the electrolyte but also electrode structure collapse, electrode/electrolyte separation, and significant polarization. Thus, the energy density and life span of aqueous lithium-ion batteries are significantly reduced. The pH of aqueous electrolyte is the primary factor that influences both potentials and reaction pathways of OER and HER (OER in acidic solution is $2H_2O \rightarrow O_2 + 4H^+ + 4e^-$ while in alkaline solution is $4OH^- \rightarrow O_2 + 2H_2O + 4e^-$; HER in acidic solution is $2H^+ + 2e^- \rightarrow H_2$ while in alkaline solution is $2H_2O + 2e^- \rightarrow H_2 + 2OH^-$).

The reaction mechanisms of aqueous lithium-ion batteries depend on electrolyte parameters such as type, pH, and concentration. LMO showed a prominent reduction peak at about 0.5 V at a low pH, and the oxidation peak slightly overlapped with that of water oxidation. While at a high pH level, the reduction peak of LMO disappeared and the oxidation peak of water increased significantly due to the gradual decrease of O_2 evolution potential in an alkaline environment [41]. When different electrolytes such as LiCl, KCl, and NH_4Cl were used, Li^+ could be extracted and inserted into LMO in LiCl electrolyte while H^+ was preferred to be inserted in KCl and NH_4Cl electrolytes [7]. Compared to the electrochemical performances of aqueous lithium-ion batteries with 1, 5, and 9 M $LiNO_3$ electrolytes, the rate capability was improved in higher concentrations of $LiNO_3$ due to the higher ionic conductivity while reversibility was best in 5 M $LiNO_3$ [8]. In addition, some electrode materials showed side reactions with electrolytes, and the geometric and electronic structures of electrode surfaces also affected the reaction mechanisms of aqueous lithium-ion batteries [5]. Therefore, it is necessary to choose appropriate aqueous electrolytes for different electrode materials to obtain the aqueous lithium-ion batteries with excellent performance.

4.1.4 Summary

This section introduces the representative electrode materials, the research progress of aqueous electrolytes, and possible reaction mechanisms in aqueous lithium-ion batteries. To improve the cycling stability of aqueous lithium-ion batteries, the aqueous electrolytes should be adjusted appropriately for different battery systems to avoid OER and HER reactions, H^+ intercalation, and dissolution of the electrode materials. In addition, the electrode material can be structurally designed or coated to maintain stability during cycling. The energy density of aqueous lithium-ion batteries is determined by the output voltage and can be improved using a water-in-salt electrolyte.

4.2 FLEXIBLE THIN-FILM AQUEOUS LITHIUM-ION BATTERIES

A flexible thin-film aqueous lithium-ion battery with a large volumetric energy density was obtained by increasing the mass loading of LMO and LTP [42]. The mixture of the LMO or LTP, acetylene black, and polytetrafluoroethylene (PTFE) was rolled into a film (Figure 4.2a). Next, the electrode was obtained by pressing the film onto a flexible stainless steel mesh. Finally, the aqueous lithium-ion battery was fabricated by the obtained electrode and microporous polyacrylonitrile nonwoven separator wetted by 2 M $LiNO_3$ aqueous electrolyte (Figure 4.2b). Considering O_2 evolution in the aqueous electrolyte when Li^+ ions were extracted from LMO during the charging process, the suitable mass ratio of the cathode and the anode was 1:1.2.

As shown in Figure 4.2c and d, active materials and acetylene black were mixed in electrode films. The line-shaped PTFE further enhanced the internal connection between active materials and acetylene black, which improved the flexibility of the aqueous lithium-ion battery. Even with a mass loading as high as 10 mg·cm^{-2}, the thickness of the obtained

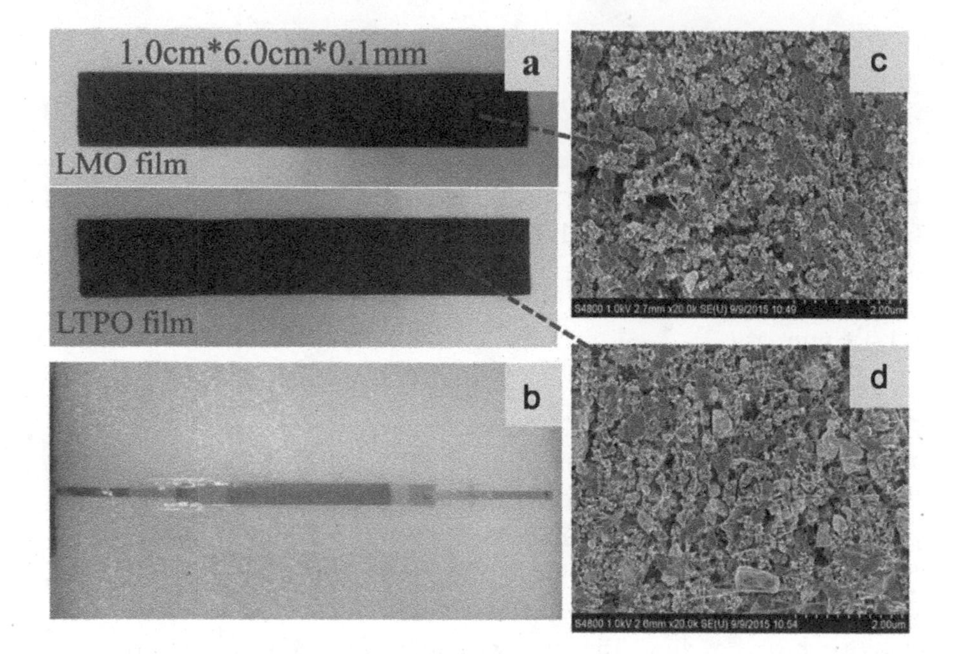

FIGURE 4.2 (a) Photographs of $LiMn_2O_4$ (LMO) and $LiTi_2(PO_4)_3$ (LTP) film electrodes. (b) Flexible thin-film aqueous lithium-ion battery. SEM images of (c) LMO and (d) LTP film electrodes. (Reproduced from Ref. [42] with permission of Wiley-VCH.)

film electrodes was 0.1 mm, and the electrode could still be freely folded, rolled, and twisted without mechanical damages.

The electrochemical performance of LMO and LTP film electrodes was tested in a 2 M $LiNO_3$ aqueous solution with a three-electrode system. Two couples of redox peaks were observed at 0.82/0.75 V and 0.95/0.88 V vs. SCE (Figure 4.3a), which corresponded to the Li^+ intercalation/deintercalation, respectively. At the current density of 1 $A{\cdot}g^{-1}$, the LMO film electrode delivered a specific capacity of 105 $mAh{\cdot}g^{-1}$, only 4% of which would lose after 100 cycles with a coulombic efficiency of ~100%. The redox peaks of the LTP film electrode at −0.81 and −0.41 V vs. SCE could be observed clearly (Figure 4.3b). The small peaks at around −0.4 V vs. SCE were to blame for the impurity of $TiPO_4$ in the LTP. The LTP film electrode exhibited a specific capacity of 80 $mAh{\cdot}g^{-1}$ at the current density of 1 $A{\cdot}g^{-1}$, which was maintained by 79% after 100 cycles with coulombic efficiency of 100%. The flexible thin-film aqueous lithium-ion battery showed a specific capacity of 40 $mAh{\cdot}g^{-1}$ at the current density of 0.2 $A{\cdot}g^{-1}$ based on the total mass of

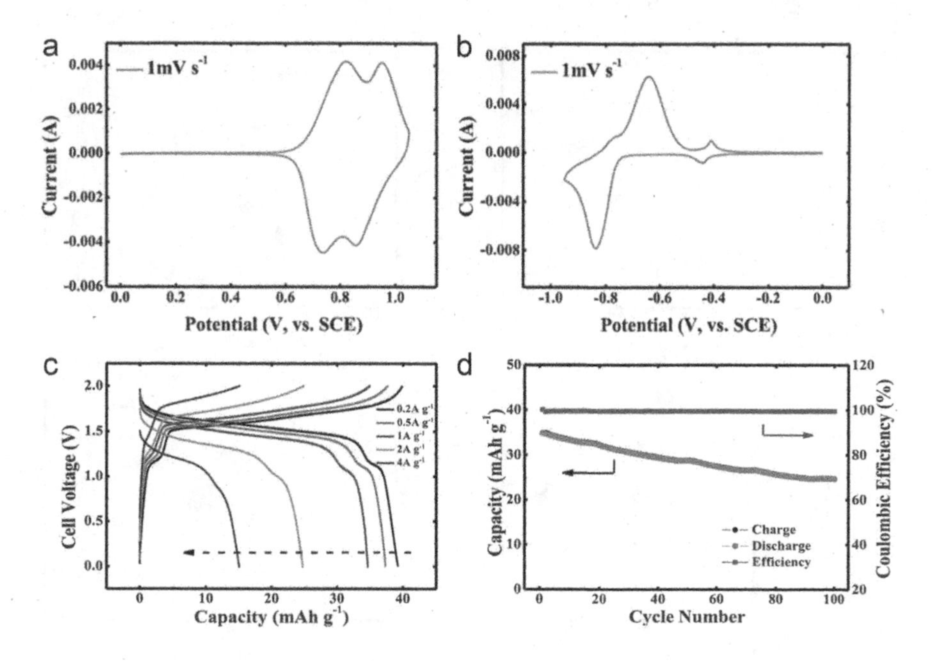

FIGURE 4.3 Cyclic voltammograms of (a) $LiMn_2O_4$ (LMO) and (b) $LiTi_2(PO_4)_3$ (LTP) film electrodes at the sweep rate of 1 $mV{\cdot}s^{-1}$. (c) Galvanostatic charge-discharge curves of the flexible thin-film aqueous lithium-ion battery at different current densities. (d) Cycle life of the flexible thin-film aqueous lithium-ion battery at the current density of 1 $A{\cdot}g^{-1}$. (Reproduced from Ref. [42] with permission of Wiley-VCH.)

electrode materials (Figure 4.3c). Furthermore, the discharge capacity at the current density of 1 $A{\cdot}g^{-1}$ was maintained at 35 $mAh{\cdot}g^{-1}$ with capacity retention of 72% during 100 cycles (Figure 4.3d). The maximum volumetric energy and power densities were 124 $mWh{\cdot}cm^{-3}$ and 11.1 $W{\cdot}cm^{-3}$, respectively, based on the total volume of the cathode, anode, and separator, which were about seven times higher than the largest values reported to date.

The thin-film aqueous lithium-ion battery exhibited high flexibility even when LMO and LTP mass loadings were about ten times higher than those previously reported. According to Figure 4.4, whether bent to 45°, 90°, 135°, 180°, or even bent into a circle, it showed almost the same electrochemical performance as the state without bending. Moreover, after 100 repeated bending, negligible performance loss was detected.

A high energy density flexible thin-film aqueous lithium-ion battery was fabricated based on $LiVPO_4F$ material as both cathode and anode and a water-in-salt gel polymer electrolyte with broad electrochemical

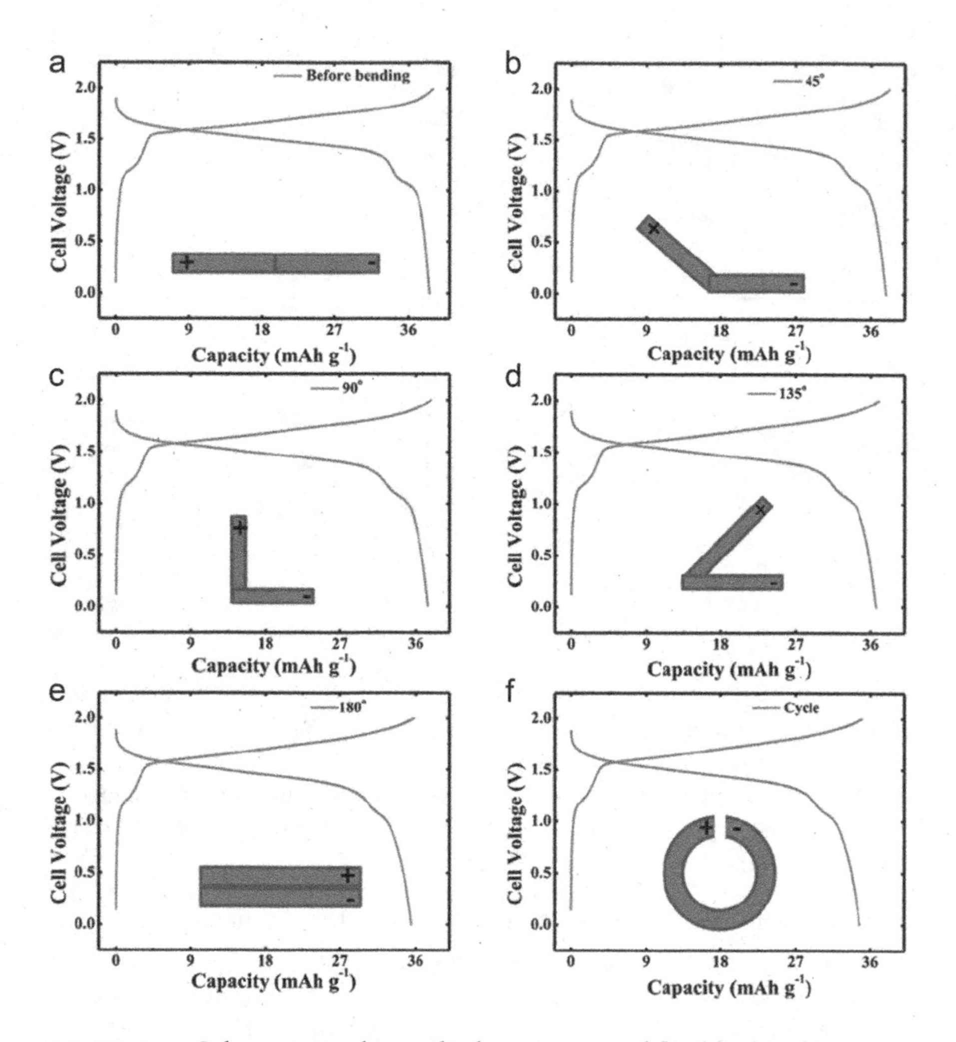

FIGURE 4.4 Galvanostatic charge-discharge curves of flexible thin-film aqueous lithium-ion battery (a) before bending, bent to (b) 45°, (c) 90°, (d) 135°, (e) 180°, and (f) bent to a cycle. (Reproduced from Ref. [42] with permission of Wiley-VCH.)

windows [43]. $LiVPO_4F$ active material was prepared through a two-step carbothermal reduction method. Then $LiVPO_4F$ was coated on the surface of Ti film, which was previously coated on the flexible Kapton film to obtain the flexible electrodes. The aqueous gel electrolytes were prepared by dissolving 25 M LiTFSI and polyvinyl alcohol in water. Finally, the flexible thin-film symmetrical aqueous lithium-ion battery was assembled by sandwiching an as-prepared thin-film water-in-salt gel electrolyte between two identical $LiVPO_4F$ electrodes through a facile lamination process.

The pomegranate-structure $LiVPO_4F$ sphere had two levels of porous structure: microsized spheres with diameters of 1–2 μm and secondary carbon-coated nanoparticles with diameters of 50–100 nm. Moreover, the highly electronic conductive three-dimensional carbon matrix can accelerate the redox reactions of $LiVPO_4F$.

$LiVPO_4F$ exhibited two couples of redox peaks at 1.98/2.18 and 4.35/4.57 V vs. Li^+/Li, which could generate an output voltage of 2.4 V from such a symmetric couple (Figure 4.5a). The lower redox peaks corresponded to

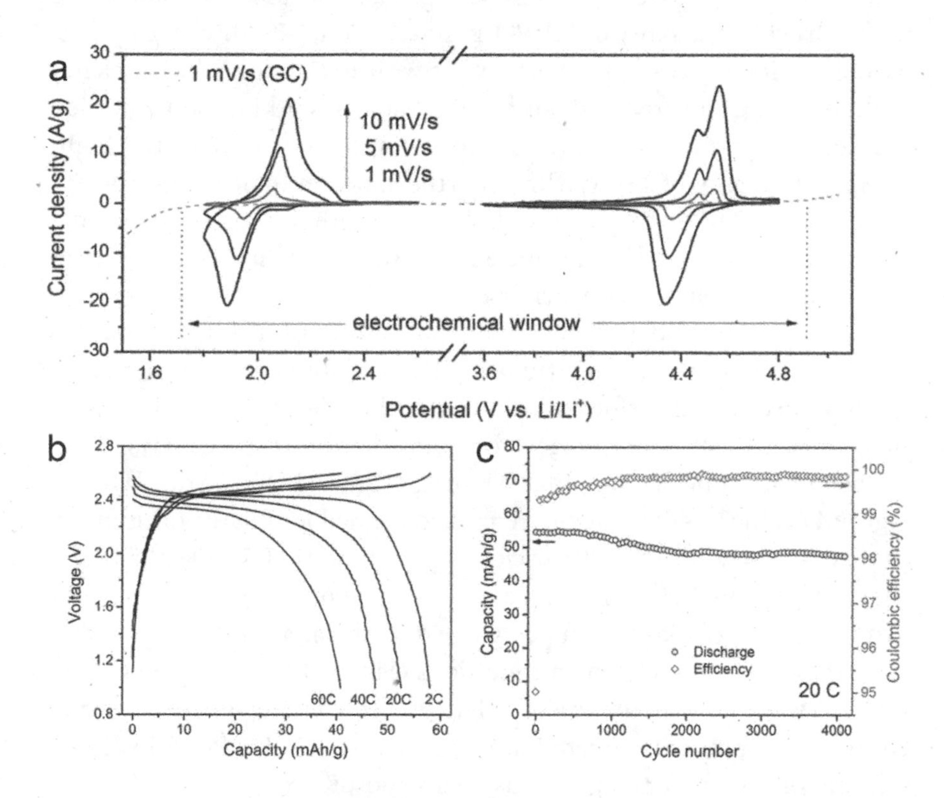

FIGURE 4.5 (a) Cyclic voltammograms of $LiVPO_4F$ samples on both cathode and anode at scan rates of 1, 5, 10 $mV \cdot s^{-1}$ and the stable electrochemical window of water-in salt gel electrolyte (dashed). (b) Galvanostatic charge-discharge curves of the symmetric $LiVPO_4F/LiVPO_4F$ aqueous lithium-ion battery at different current rates. (c) Cycle performance of the symmetric $LiVPO_4F/LiVPO_4F$ aqueous lithium-ion battery at a current rate of 20 C. (Reproduced from Ref. [43] with permission of Wiley-VCH.)

the reduction of V^{III} to V^{II} while the higher redox peaks corresponded to the oxidation of V^{III} to V^{IV}. The theoretical capacities of the two processes were both 156 $mAh{\cdot}g^{-1}$. The output voltage generated by $LiVPO_4F$ perfectly fitted the stable electrochemical window of the water-in-salt gel electrolyte (1.72–4.92 V vs. SCE).

The symmetric $LiVPO_4F/LiVPO_4F$ aqueous lithium-ion battery exhibited superior rate capability. At the current rate of 2 C (0.314 $A{\cdot}g^{-1}$), a symmetric lithium-ion battery delivered an average output voltage of 2.42 V and a discharge capacity of 58.1 $mAh{\cdot}g^{-1}$. At the rate of 60 C (9.42 $A{\cdot}g^{-1}$), it still delivered an average output voltage of 1.29 V and a discharge capacity of 40.8 $mAh{\cdot}g^{-1}$ (Figure 4.5b). The high rate capability was attributed to the structural stability of $LiVPO_4F$. In addition, 87% discharge capacity was retained at 20 C after 4,000 cycles with 100% coulombic efficiency (Figure 4.5c). Compared with all reported aqueous lithium-ion batteries, the flexible symmetric $LiVPO_4F$ lithium-ion battery showed the highest specific energy of 141 $Wh{\cdot}kg^{-1}$. Furthermore, even at a high specific power of 20,600 $W{\cdot}kg^{-1}$, the flexible $LiVPO_4F$ lithium-ion battery could still deliver 90 $Wh{\cdot}kg^{-1}$ of specific energy, which was much higher than most of the aqueous supercapacitors.

It should be noted that the symmetric $LiVPO_4F/LiVPO_4F$ aqueous lithium-ion battery exhibited tremendous flexibility and intrinsic safety against mechanical deformation under the electrochemical cycling state (Figure 4.6a). The voltage profiles of the first two scenarios were almost identical before and after bending (Figure 4.6b). As displayed in Figure 4.6c, the loss of capacities remained negligible after the lithium-ion battery was bent for over 200 cycles. The symmetric $LiVPO_4F/LiVPO_4F$ aqueous lithium-ion battery, with a capacity of about 60 mAh, only exhibited a slight energy loss when being cut in ambient air and continued to drive the small fan under open conditions (Figure 4.6d). The strong bond between water and salt in the electrolyte prevented water evaporation and ensured the continuous operation of the symmetric $LiVPO_4F/LiVPO_4F$ lithium-ion battery for long periods after exposure to air.

4.3 FLEXIBLE FIBER AQUEOUS LITHIUM-ION BATTERIES

Although certain progress in achieving flexibility is made in thin-film flexible aqueous lithium-ion batteries, their flexibility is still restricted by their two-dimensional nature. Compared with the thin-film structure,

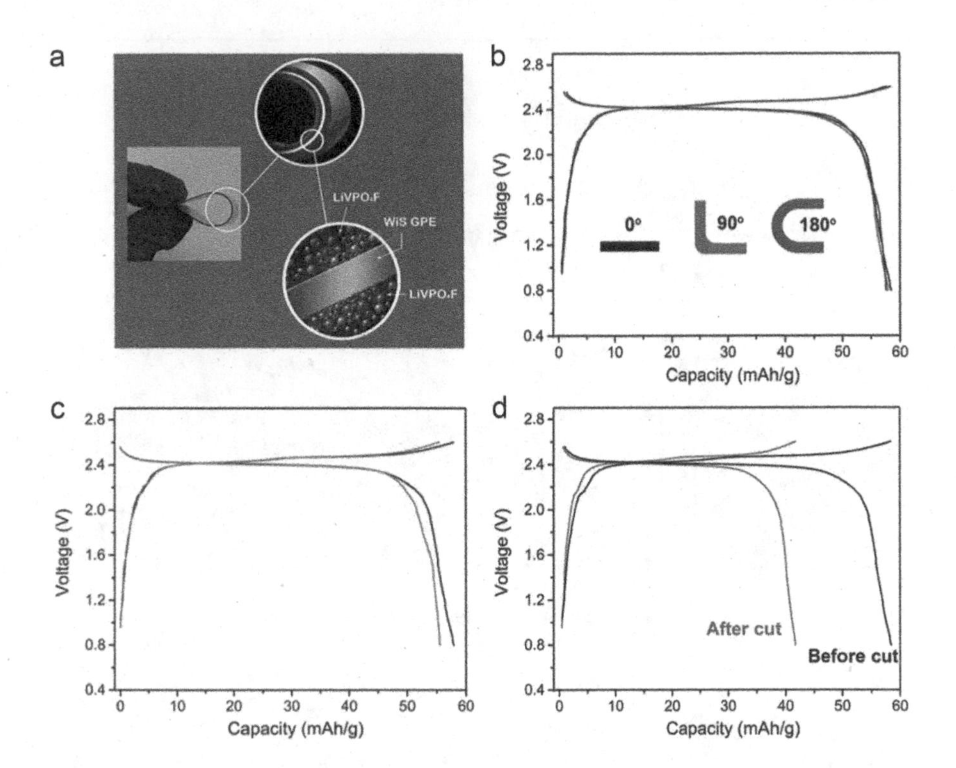

FIGURE 4.6 (a) Photograph of the fabricated symmetric $LiVPO_4F/LiVPO_4F$ aqueous lithium-ion battery and its flexibility. Galvanostatic charge-discharge curves of the symmetric $LiVPO_4F/LiVPO_4F$ aqueous lithium-ion battery at 2 C (b) under different bending angles, (c) after 200 cycles of bending, (d) before and after cutting. (Reproduced from Ref. [43] with permission of Wiley-VCH.)

one-dimensional fiber batteries can bear deformations in all directions. Besides, the fiber batteries can be woven into clothes, which is attractive for wearable electronics.

A flexible fiber aqueous lithium-ion battery with high power density was fabricated by a polyimide (PI)/carbon nanotube (CNT) fiber anode, an LMO/CNT fiber cathode, and a Li_2SO_4 aqueous electrolyte [44]. PI/CNT hybrid fiber electrode was synthesized by *in situ* polymerization of PI on aligned CNT fiber. To prepare LMO/CNT hybrid fiber electrode, the LMO suspension was first dropped on the CNT sheets, followed by twisting into a hybrid fiber. The fiber aqueous lithium-ion battery was fabricated by inserting the above two hybrid fibers into a heat-shrinkable tube with a separator, followed by injecting Li_2SO_4 solution into the tube and finally sealing with a heat gun (Figure 4.7a).

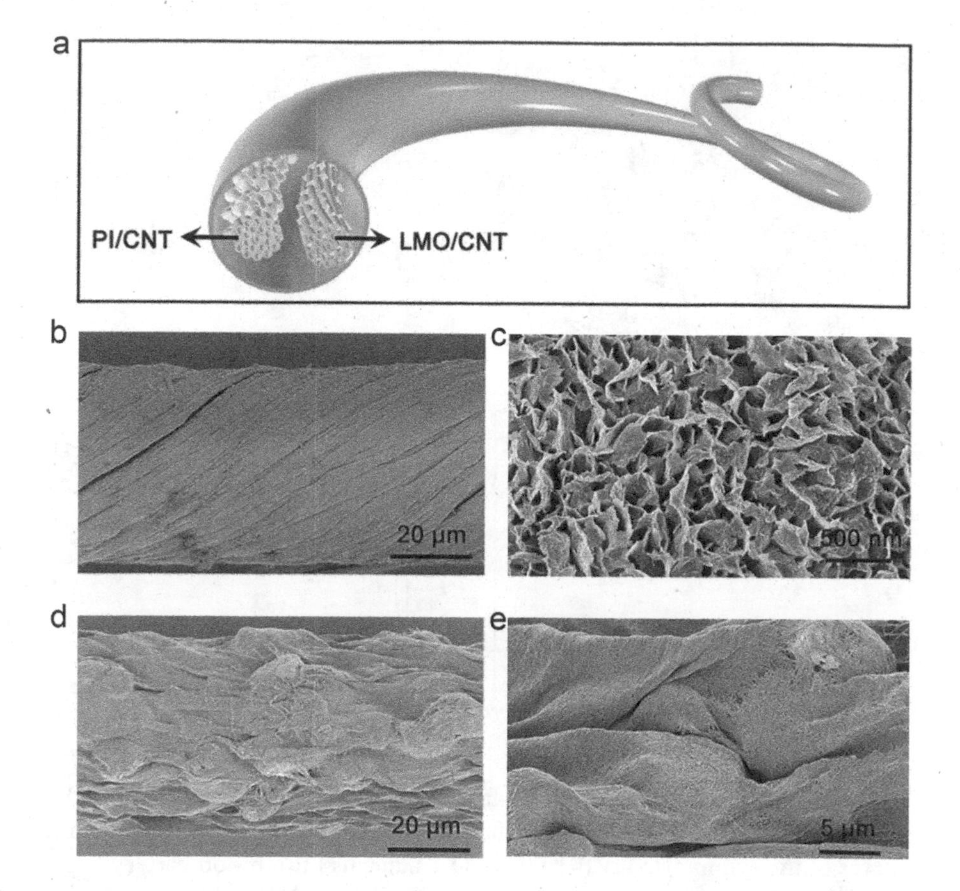

FIGURE 4.7 (a) Schematic illustration of the structure of the fiber aqueous lithium-ion battery. (b and c) SEM images of polyimide (PI)/carbon nanotube (CNT) hybrid fiber at low and high magnifications, respectively. (d and e) SEM images of $LiMn_2O_4$ (LMO)/CNT hybrid fiber at low and high magnifications, respectively. (Reproduced from Ref. [44] with permission of the Royal Society of Chemistry.)

PI/CNT hybrid fiber electrode exhibited a skin-core structure. The internally aligned CNT fiber's aligned structure and mechanical properties were well maintained with a slight increase in diameter (Figure 4.7b). PI nanosheets with widths of ~300 nm and thicknesses of ~25 nm were uniformly distributed vertically on the CNT fibers' surface, and many voids were produced between the sheets (Figure 4.7c). The loading content of PI was 53% in the hybrid fiber. For LMO/CNT hybrid fiber electrode, the spinel LMO nanoparticles with an average diameter of ~200 nm were well wrapped by CNTs through a co-spinning process (Figure 4.7d and e). The loading content of LMO was

49%. The PI/CNT and LMO/CNT hybrid fibers were highly flexible. After bending for 1,000 cycles, no apparent structural damage was observed.

The electrochemical performances of the two hybrid fibers were tested in a standard three-electrode system in 0.5 M Li_2SO_4. The redox potentials of PI/CNT were observed at around −0.8 and −0.6 V vs. SCE. The PI/CNT hybrid fiber electrode delivered a specific capacity of 129.8 $mAh \cdot g^{-1}$ at a rate of 20 C and retained 86 $mAh \cdot g^{-1}$ at 600 C (Figure 4.8a). The high rate performance was attributed to intrinsically insulating PI and highly electronic conductive aligned CNT fiber. It also exhibited a stable cycling performance, as shown in Figure 4.8b. More than 90% of the specific capacity was maintained after 200 cycles at the rate of 20 C with a coulombic efficiency of ~100%. The redox potentials of LMO/CNT were around 0.7 and 0.8 V vs. SCE. The discharge capacities of the LMO/CNT hybrid fiber electrode were 140 $mAh \cdot g^{-1}$ at the rate of 10 C, and they were maintained by 84.7 $mAh \cdot g^{-1}$ at 100 C (Figure 4.8c). Such an outstanding rate performance

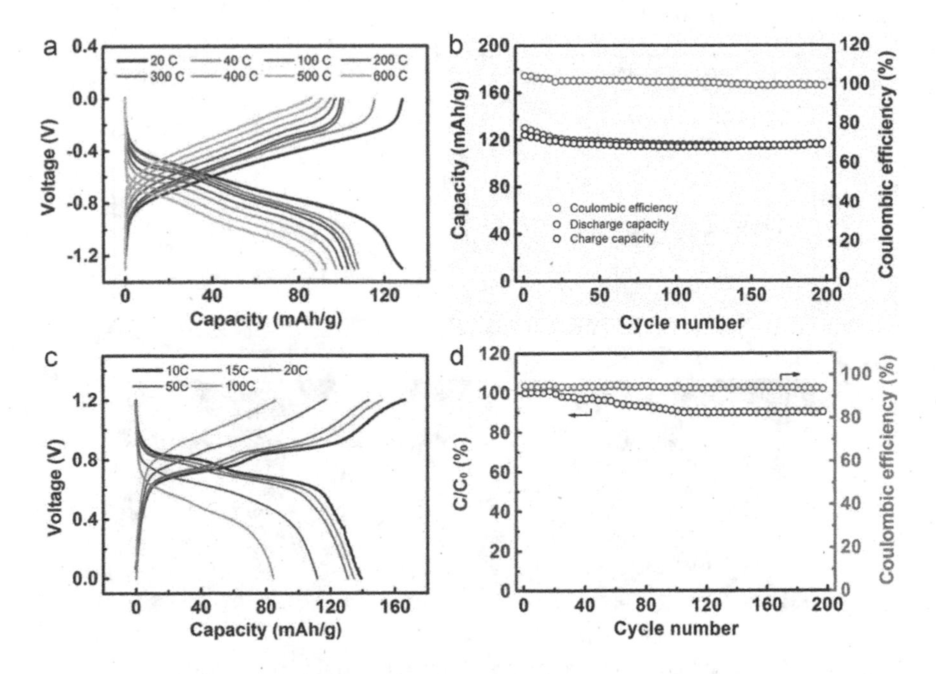

FIGURE 4.8 (a) Galvanostatic charge-discharge curves and (b) long-term stability test of polyimide (PI)/carbon nanotube (CNT) hybrid fiber. (c) Galvanostatic charge-discharge curves and (d) long-term stability test of $LiMn_2O_4$ (LMO)/CNT hybrid fiber. (Reproduced from Ref. [44] with permission of the Royal Society of Chemistry.)

was derived from the nanostructure of LMO, which shortened the transport distance of Li^+ efficiently. According to Figure 4.8d, LMO/CNT also exhibited high cycling stability. It could maintain over 90% of the specific capacity after 200 cycles at a current rate of 10 C.

The fiber aqueous lithium-ion battery showed a discharge potential platform of 1.4 V at the current rate of 10 C. The specific discharge capacity of the battery was 123 mAh·g^{-1} and the specific capacity was 101 mAh·g^{-1} even at a current rate of 100 C. Furthermore, the specific capacity was well maintained with a coulombic efficiency of around 98% after over 1,000 cycles (Figure 4.9a and b). With the consideration of the mass of PI, LMO, CNT, electrolyte,

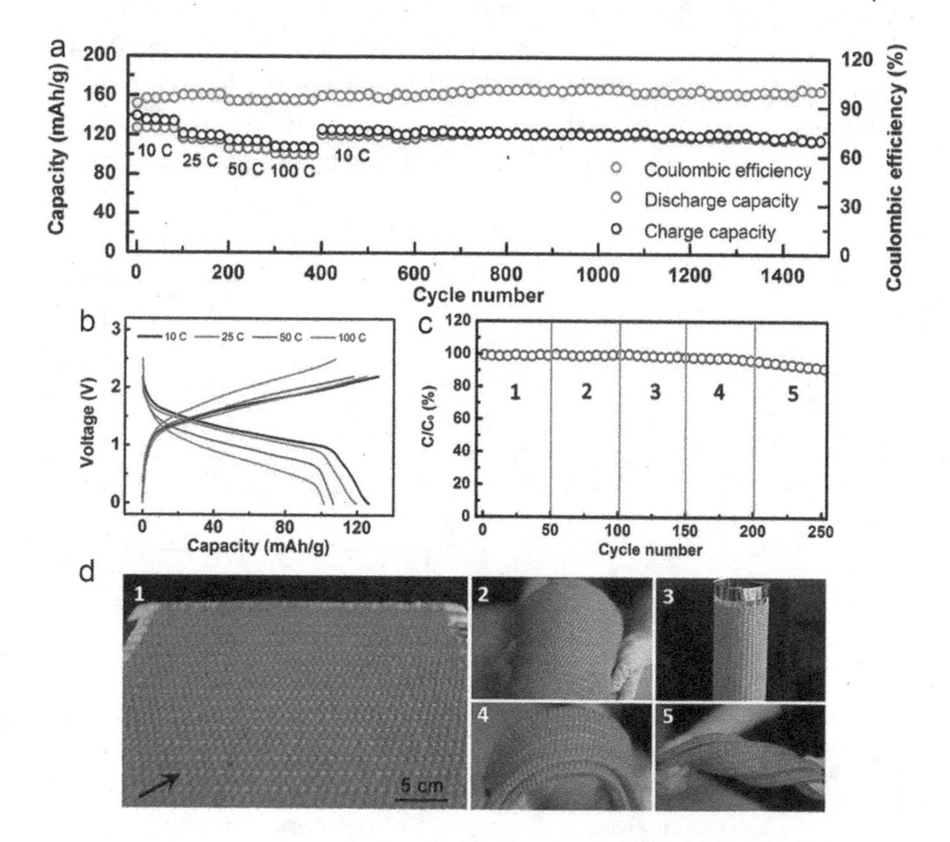

FIGURE 4.9 (a) Rate performance and long-term stability performance of fiber aqueous lithium-ion battery. (b) Charge-discharge curves at different current rates. (c) C_0 and C correspond to the capacities before and after the deformation of the cloth, respectively. (d) Cloth woven from the fiber aqueous lithium-ion battery and under bending, folding, and twisting in order. (Reproduced from Ref. [44] with permission of the Royal Society of Chemistry.)

and package materials, the power density and energy density of the full cell were 3,984.9 W kg^{-1} and 19.1 W h kg^{-1}, respectively, which were 10–20 times of the previous fiber-shaped organic lithium-ion battery. This fiber aqueous lithium-ion battery was able to maintain its good electrochemical performance after being woven into clothes (Figure 4.9c), which could be bent, folded, and twisted into different shapes (Figure 4.9d). This proves the advantage of fiber aqueous lithium-ion batteries for wearable electronic products.

4.4 PERSPECTIVE

With the rapid development of wearable devices, higher requirements for safe batteries have been proposed. Traditional organic lithium-ion batteries are restricted by significant safety and health hazards when a competitive successor, aqueous lithium-ion batteries, fills the gap. The primary difference lies in the transition from organic to aqueous electrolytes, bringing inherent OER and HER reactions and a narrower working window. To realize high-performance aqueous lithium-ion batteries, it is vital to match the electrolytes' working windows with the electrodes' reaction potentials. Hence, many flexible aqueous lithium-ion batteries have been developed while explorations have been made on their flexibility and configuration.

However, the large-scale application of aqueous lithium-ion batteries in wearable devices is still rugged. First, the use of aqueous electrolytes inevitably leads to a loss of energy density and life span. Despite achievements in high-voltage window aqueous electrolytes, the electrochemical performance is still far inferior to organic lithium-ion batteries. Moreover, the available aqueous electrolytes with high-voltage windows are liquid, which is difficult to conform to wearable devices. The current designation of aqueous gel electrolytes is still based on research on aqueous liquid electrolytes, and there are few studies on aqueous gel electrolytes with high-voltage windows. The new host materials or chemistries with more electron transfer can be adopted to improve the energy density for electrode materials. The pseudocapacitive behavior is more favorable for the high rate and power density than the volumetric lattice diffusion. In addition, the cost is also needed to be considered for commercialization.

REFERENCES

1. Kang, K. Meng, Y. S. Breger, J. Grey, C. P. Ceder, G. 2006. Electrodes with high power and high capacity for rechargeable lithium batteries. *Science* 311: 977.

2. Winter, M. Besenhard, J. O. Spahr, M. E. Novµk, P. 1998. Insertion electrode materials for rechargeable lithium batteries. *Advanced Materials* 10: 725.
3. Ye, T. Li, L. Zhang, Y. 2020. Recent progress in solid electrolytes for energy storage devices. *Advanced Functional Materials* 30: 2000077.
4. Kim, H. Hong, J. Park, K. Y. Kim, H. Kim, S. W. Kang, K. 2014. Aqueous rechargeable Li and Na ion batteries. *Chemical Reviews* 114: 11788–11827.
5. Chao, D. Zhou, W. Xie, F. Ye, C. Li, H. Jaroniec, M. Qiao, S.-Z. 2020. Roadmap for advanced aqueous batteries: from design of materials to applications. *Science Advances* 6: eaba4098.
6. Li, W. Dahn, J. R. Wainwright, D. S. 1994. Rechargeable lithium batteries with aqueous electrolytes. *Science* 264: 1115–1118.
7. Jayalakshmi, M. Rao, M. M. Scholz, F. 2003. Electrochemical behavior of solid lithium manganate ($LiMn_2O_4$) in aqueous neutral electrolyte solutions. *Langmuir* 19: 8403–8408.
8. Tian, L. Yuan, A. 2009. Electrochemical performance of nanostructured spinel $LiMn_2O_4$ in different aqueous electrolytes. *Journal of Power Sources* 192: 693–697.
9. Yuan, A. Tian, L. Xu, W. Wang, Y. 2010. Al-doped spinel $LiAl_{0.1}Mn_{1.9}O_4$ with improved high-rate cyclability in aqueous electrolyte. *Journal of Power Sources* 195: 5032–5038.
10. Cvjeticanin, N. Stojkovic, I. Mitric, M. Mentus, S. 2007. Cyclic voltammetry of $LiCr_{0.15}Mn_{1.85}O_4$ in an aqueous $LiNO_3$ solution. *Journal of Power Sources* 174: 1117–1120.
11. Tang, W. Hou, Y. Wang, F. Liu, L. Wu, Y. Zhu, K. 2013. $LiMn_2O_4$ nanotube as cathode material of second-level charge capability for aqueous rechargeable batteries. *Nano Letter* 13: 2036–2040.
12. Qu, Q. Fu, L. Zhan, X. Samuelis, D. Maier, J. Li, L. Tian, S. Li, Z. Wu, Y. 2011. Porous $LiMn_2O_4$ as cathode material with high power and excellent cycling for aqueous rechargeable lithium batteries. *Energy & Environmental Science* 4: 3985–3990.
13. Ruffo, R. Wessells, C. Huggins, R. A. Cui, Y. 2009. Electrochemical behavior of $LiCoO_2$ as aqueous lithium-ion battery electrodes. *Electrochemistry Communications* 11: 247–249.
14. Wang, G. J. Qu, Q. T. Wang, B. Shi, Y. Tian, S. Wu, Y. P. Holze, R. 2009. Electrochemical behavior of $LiCoO_2$ in a saturated aqueous Li_2SO_4 solution. *Electrochimica Acta* 54: 1199–1203.
15. Tang, W. Liu, L. L. Tian, S. Li, L. Yue, Y. B. Wu, Y. P. Guan, S. Y. Zhu, K. 2010. Nano-$LiCoO_2$ as cathode material of large capacity and high rate capability for aqueous rechargeable lithium batteries. *Electrochemistry Communications* 12: 1524–1526.
16. He, P. Liu, J.-L. Cui, W.-J. Luo, J.-Y. Xia, Y.-Y. 2011. Investigation on capacity fading of $LiFePO_4$ in aqueous electrolyte. *Electrochimica Acta* 56: 2351–2357.

17. Murphy, D. W. Christian, P. A. DiSalvo, F. J. Carides, J. N. Waszczak, J. V. 1981. Lithium incorporation by V_6O_{13} and related vanadium (+4, +5) oxide cathode materials. *Journal of the Electrochemical Society* 128: 2053.
18. Ni, J. Jiang, W. Yu, K. Gao, Y. Zhu, Z. 2011. Hydrothermal synthesis of VO_2(b) nanostructures and application in aqueous Li-ion battery. *Electrochimica Acta* 56: 2122–2126.
19. Wang, F. Liu, Y. Liu, C.-Y. 2010. Hydrothermal synthesis of carbon/vanadium dioxide core-shell microspheres with good cycling performance in both organic and aqueous electrolytes. *Electrochimica Acta* 55: 2662–2666.
20. Kohler, J. Makihara, H. Uegaito, H. Inoue, H. Toki, M. 2000. Liv3o8: characterization as anode material for an aqueous rechargeable Li-ion battery system. *Electrochimica Acta* 46: 59–65.
21. Caballero, A. Morales, J. Vargas, O. A. 2010. Electrochemical instability of LiV_3O_8 as an electrode material for aqueous rechargeable lithium batteries. *Journal of Power Sources* 195: 4318–4321.
22. Cheng, C. Li, Z. H. Zhan, X. Y. Xiao, Q. Z. Lei, G. T. Zhou, X. D. 2010. A macaroni-like $Li_{1.2}V_3O_8$ nanomaterial with high capacity for aqueous rechargeable lithium batteries. *Electrochimica Acta* 55: 4627–4631.
23. Zhao, M. Zhang, B. Huang, G. Zhang, H. Song, X. 2013. Excellent rate capabilities of ($LiFePO_4$/C)//LiV_3O_8 in an optimized aqueous solution electrolyte. *Journal of Power Sources* 232: 181–186.
24. Whittingham, M. S. 2004. Lithium batteries and cathode materials. *Chemical Reviews* 10: 4271–4302.
25. Wang, H. Zeng, Y. Huang, K. Liu, S. Chen, L. 2007. Improvement of cycle performance of lithium ion cell $LiMn_2O_4$/LixV_2O_5 with aqueous solution electrolyte by polypyrrole coating on anode. *Electrochimica Acta* 52: 5102–5107.
26. Wang, H. Huang, K. Zeng, Y. Zhao, F. Chen, L. 2007. Stabilizing cyclability of an aqueous lithium-ion battery $LiNi_{1/3}Mn_{1/3}Co_{1/3}O_2/Li_xV_2O_5$ by polyaniline coating on the anode. *Electrochemical and Solid-State Letters* 10: A199.
27. Ibrahim, H. Ilinca, A. Perron, J. 2008. Energy storage systems-characteristics and comparisons. *Renewable and Sustainable Energy Reviews* 12: 1221–1250.
28. Luo, J. Y. Xia, Y. Y. 2007. Aqueous lithium-ion battery $LiTi_2(PO_4)_3/LiMn_2O_4$ with high power and energy densities as well as superior cycling stability. *Advanced Functional Materials* 17: 3877–3884.
29. Wessells, C. La Mantia, F. Deshazer, H. Huggins, R. A. Cui, Y. 2011. Synthesis and electrochemical performance of a lithium titanium phosphate anode for aqueous lithium-ion batteries. *Journal of the Electrochemical Society* 158: A352.
30. Armand, M. Grugeon, S. Vezin, H. Laruelle, S. Ribiere, P. Poizot, P. Tarascon, J. M. 2009. Conjugated dicarboxylate anodes for li-ion batteries. *Nature Materials* 8: 120–125.

31. Walker, W. Grugeon, S. Mentre, O. Laruelle, S. P. Tarascon, J.-M. Wudl, F. 2010. Ethoxycarbonyl-based organic electrode for Li-batteries. *Journal of the American Chemical Society* 132: 6517–6523.
32. Wang, G. Qu, Q. Wang, B. Shi, Y. Tian, S. Wu, Y. 2008. An aqueous electrochemical energy storage system based on doping and intercalation: Ppy//$LiMn_2O_4$. *Chemphyschem* 9: 2299–2301.
33. Qin, H. Song, Z. P. Zhan, H. Zhou, Y. H. 2014. Aqueous rechargeable alkali-ion batteries with polyimide anode. *Journal of Power Sources* 249: 367–372.
34. Suo, L. Borodin, O. Gao, T. Olguin, M. Ho, J. Fan, X. Luo, C. Wang, C. Xu, K. 2015. "Water-in-salt" electrolyte enables high-voltage aqueous lithium-ion chemistries. *Science* 350: 938–943.
35. Suo, L. Borodin, O. Sun, W. Fan, X. Yang, C. Wang, F. Gao, T. Ma, Z. Schroeder, M. von Cresce, A. Russell, S. M. Armand, M. Angell, A. Xu, K. Wang, C. 2016. Advanced high-voltage aqueous lithium-ion battery enabled by "water-in-bisalt" electrolyte. *Angewandte Chemie* 128: 7252–7257.
36. Yamada, Y. Usui, K. Sodeyama, K. Ko, S. Tateyama, Y. Yamada, A. 2016. Hydrate-melt electrolytes for high-energy-density aqueous batteries. *Nature Energy* 1: 1–9.
37. Ko, S. Yamada, Y. Miyazaki, K. Shimada, T. Watanabe, E. Tateyama, Y. Kamiya, T. Honda, T. Akikusa, J. Yamada, A. 2019. Lithium-salt monohydrate melt: a stable electrolyte for aqueous lithium-ion batteries. *Electrochemistry Communications* 104: 106488.
38. Lukatskaya, M. R. Feldblyum, J. I. Mackanic, D. G. Lissel, F. Michels, D. L. Cui, Y. Bao, Z. 2018. Concentrated mixed cation acetate "water-in-salt" solutions as green and low-cost high voltage electrolytes for aqueous batteries. *Energy & Environmental Science* 11: 2876–2883.
39. Xie, J. Liang, Z. Lu, Y. C. 2020. Molecular crowding electrolytes for high-voltage aqueous batteries. *Nature Materials* 19: 1006–1011.
40. He, X. Yan, B. Zhang, X. Liu, Z. Bresser, D. Wang, J. Wang, R. Cao, X. Su, Y. Jia, H. Grey, C. P. Frielinghaus, H. Truhlar, D. G. Winter, M. Li, J. Paillard, E. 2018. Fluorine-free water-in-ionomer electrolytes for sustainable lithium-ion batteries. *Nature Communication* 9: 5320.
41. Pei, W. Hui, Y. Huaquan, Y. 1996. Electrochemical behavior of Li-Mn spinel electrode material in aqueous solution. *Journal of Power Sources* 63: 275–278.
42. Dong, X. Chen, L. Su, X. Wang, Y. Xia, Y. 2016. Flexible aqueous lithium-ion battery with high safety and large volumetric energy density. *Angewandte Chemie* 128: 7600–7603.
43. Yang, C. Ji, X. Fan, X. Gao, T. Suo, L. Wang, F. Sun, W. Chen, J. Chen, L. Han, F. Miao, L. Xu, K. Gerasopoulos, K. Wang, C. 2017. Flexible aqueous Li-ion battery with high energy and power densities. *Advanced Materials* 29: 1701972.
44. Zhang, Y. Wang, Y. Wang, L. Lo, C.-M. Zhao, Y. Jiao, Y. Zheng, G. Peng, H. 2016. A fiber-shaped aqueous lithium ion battery with high power density. *Journal of Materials Chemistry A* 4: 9002–9008.

CHAPTER 5

Flexible Aqueous Sodium-Ion Batteries

5.1 OVERVIEW OF AQUEOUS SODIUM-ION BATTERIES

Lithium-ion batteries have been widely studied and successfully applied in various electronic devices in our daily lives. However, intrinsic cost and safety characteristics hinder the development of current lithium-ion batteries [1]. On the one hand, the cost of lithium-ion batteries is relatively high due to the high price of lithium salts, transition metals, organic electrolytes, and the high standard of the dry environment during manufacturing processes. On the other hand, the high safety risks of lithium-ion batteries derive from the toxic and flammable organic electrolytes and the thermal runaway caused by the reactivity of the electrode materials with electrolytes.

As the resource of sodium (2.83% in the earth's crust) is far more abundant than lithium (20 ppm in the earth's crust), sodium-ion batteries are attracting extensive attention as a promising alternative candidate of lithium-ion batteries in recent years [2,3]. Owing to the similarity of the operation mechanism and fabrication technologies (such as current collector and organic electrolyte) to conventional lithium-ion batteries, a similar production line can be directly used to develop sodium-ion batteries. However, traditional sodium-ion batteries are based on organic electrolytes, which use highly toxic and flammable organic solvents and lead

DOI: 10.1201/9781003273677-5

to severe safety hazards [2,4]. Rechargeable aqueous sodium-ion batteries are endowed with lower cost and higher safety, which are promising alternatives [5,6].

5.1.1 Working Mechanism

Sodium-ion and lithium-ion batteries' working mechanisms and battery components are the same except for their ion carriers. Sodium ions diffuse through the electrolyte in sodium-ion batteries, and electrons move through the external circuit between the cathode and the anode. In particular, the working principle of aqueous sodium-ion batteries is quite similar to that of conventional nonaqueous sodium-ion batteries (Figure 5.1). The radius of sodium ions (0.102 nm) is more significant than that of lithium ions, making the insertion reaction of sodium ions more complicated

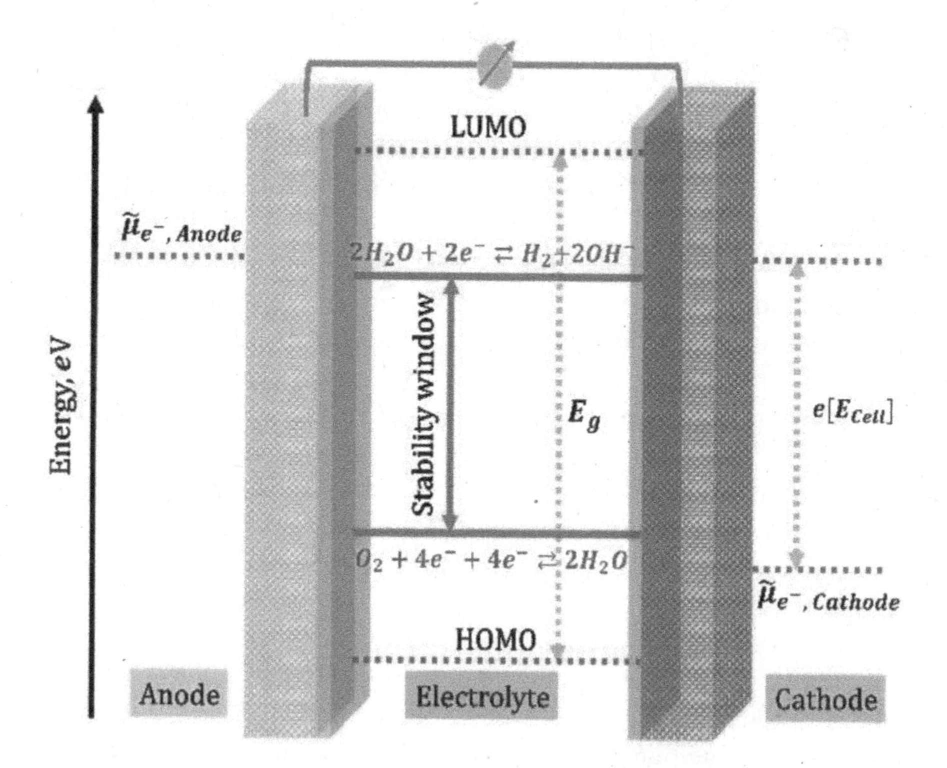

FIGURE 5.1 Schematic diagram of the working mechanism of aqueous sodium-ion batteries (HOMO, highest occupied molecular orbital and LUMO, lowest unoccupied molecular orbital). (Reproduced from Ref. [7] with permission of the American Chemical Society.)

than the insertion reaction of lithium ions. In addition, the large radius of sodium ions makes the collapse of electrode structures much easier and affects the cycling stability of the aqueous sodium-ion batteries

The aqueous electrolytes support the migration of ions in aqueous sodium-ion batteries. The operation of the electrolyte is limited by the energy difference between the highest occupied molecular orbital (HOMO) and the lowest unoccupied molecular orbital (LUMO) [7]. The organic liquid electrolytes usually undergo degradation reactions to form solid electrolyte interphases (SEIs) on the surface of electrodes, which can facilitate the diffusion process of charge carriers. However, SEI generally was not formed in aqueous sodium-ion batteries. Gaseous H_2 and O_2 might evolve from the aqueous electrolytes because the thermodynamic electrochemical stability window of water is only 1.23 V. Even combining dynamics, the electrochemical stability windows of aqueous electrolytes are still usually no more than 1.8 V, which brings many challenges for aqueous sodium-ion batteries, including the relatively low energy density, limited choice of electrode materials, and side reactions between electrode materials and water/O_2 that deteriorate the cycling stability of the battery [8,9].

5.1.2 Electrode Active Materials

Although the working mechanism of aqueous sodium-ion batteries is quite similar to conventional nonaqueous sodium-ion batteries, the extraction and insertion of sodium ions in aqueous electrolytes are more complicated, which dramatically influences the selection of electrode materials for aqueous sodium-ion batteries [10]. To avoid the side reactions of H_2 and O_2 evolutions, the redox potentials of cathodes should be lower than the oxygen evolution potential and that of the anodes should be higher than the hydrogen evolution potential in the electrolytes. Therefore, most of the conventional high-potential cathodes and low-potential anodes are not suitable for aqueous sodium-ion batteries.

The fundamental working conditions of aqueous sodium-ion batteries depend on the intercalation of sodium ions into electrode active materials in the aqueous environment. The side reactions between electrode materials and H_2O or residual O_2 will tremendously deteriorate the cycling stability of the aqueous batteries [11]. In addition, the dissolution of electrode materials might be involved in aqueous electrolytes. Therefore, the electrode materials of aqueous sodium-ion batteries should be chemically

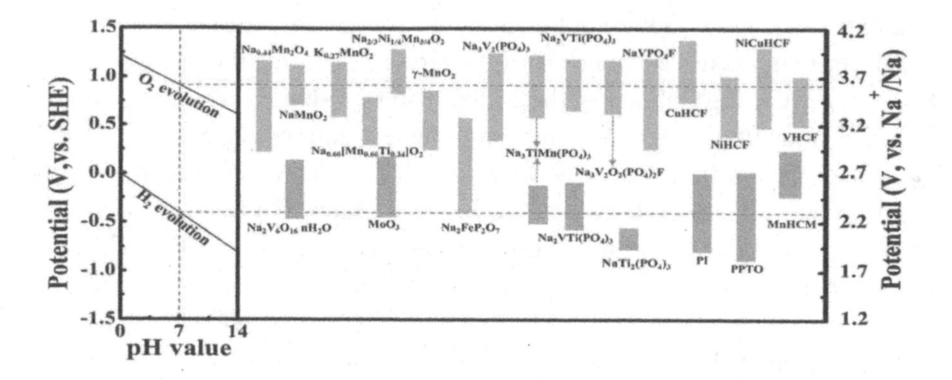

FIGURE 5.2 The redox potential of the electrode materials for aqueous sodium-ion batteries and the electrochemical stability window of aqueous electrolyte with different pH values. (Reproduced from Ref. [10] with permission of Wiley.)

stable in a specific pH range to avoid the dissolution of electrodes in aqueous electrolytes. The redox potential of the cathode and anode materials for aqueous sodium-ion batteries are shown in Figure 5.2.

For cathode materials, oxides (such as $Na_{0.44}MnO_2$ [12] and MnO_2 [13,14]), polyanionic compounds (such as $Na_2FeP_2O_7$ [15], NASICON-Type $Na_3V_2(PO_4)_3$ [16], and $NaVPO_4F$ [17]), Prussian blue analogues ($A_xPR(CN)_6$ [18]), and other Na-based compounds [19] have been developed for aqueous sodium-ion batteries in recent years. For anode materials, polyanionic compounds (such as NASICON-Type $NaTi_2(PO_4)_3$ [20], $Na_2VTi(PO_4)_3$ [21], and $Na_3MgTi(PO_4)_3$ [22]), Prussian blue analogues (such as manganese hex-acyanometallate), oxides (such as $V_2O_5{\cdot}0.6H_2O$ [23], $Na_2V_6O_{16}{\cdot}nH_2O$ [24], and $NaV_3(PO_4)_3$ [25]), and other compounds (such as MoO_3 [26], poly(2-vinylan-thraquinone) [27] and 1,4,5,8-naphthalenetetracarbox ylic dianhydride-derived polyimide) have been developed for aqueous sodium-ion batteries.

The stability of electrodes of aqueous sodium-ion batteries in aqueous electrolytes is critical since sodium ions are quite reactive in aqueous solutions. The sustainability of the electrode in the electrolyte solutions is dependent on the pH of the solution. Most of the electrodes are stable in a specific potential range at neutral pH. Surface modification of the electrodes effectively avoids the undesired dissolution of electrodes in the electrolyte and improves the electrochemical performance. In addition, surface-protected composite electrodes obtained from pyrolysis of carbon from carbon sources such as glucose and sucrose can also suppress the electrode dissolution process [28].

5.1.3 Aqueous Sodium-Ion Electrolytes

For conventional sodium-ion batteries, the low ionic conductivities in nonaqueous electrolytes have limited their electrochemical performances. Aqueous electrolytes are safer and cheaper than organic ones, which can overcome the fundamental issues caused by organic electrolytes [29]. Although the electrochemical stability windows of aqueous electrolytes are relatively narrow, they are environmentally green, which can be easily handled during the fabrication process of the battery under ambient conditions. In addition, aqueous electrolytes have higher ionic conductivities of an order magnitude than nonaqueous electrolytes, which makes it possible to achieve high power densities for aqueous batteries [30].

Different aqueous electrolytes have been extensively used in aqueous sodium-ion batteries. The electrochemical stability windows of the aqueous electrolytes can be investigated by linear sweep voltammetry, where a stainless steel or titanium mesh is used as the working electrode in a three-electrode configuration. The commonly used sodium salts in aqueous sodium-ion electrolytes are Na_2SO_4 [31,32], $NaNO_3$ [17], $NaClO_4$ [33,34], and NaCl [35]. The most-reported electrolyte for aqueous sodium-ion batteries is 1 M Na_2SO_4 with neutral pH, which is compatible with various electrode materials, including $Na_{0.44}MnO_2$ [36], $NaTi_2(PO_4)_3$ [37,38], $Na_3V_2(PO_4)_3$ [39], and $Na_3MnTi(PO_4)_3$ [40].

The electrochemical stability windows of aqueous electrolytes are strongly related to the concentrations of the salts in electrolytes. A higher concentration of electrolytes can enable a wider operating potential window for the electrolyte. For example, the electrochemical stability window of $NaClO_4$ aqueous electrolyte increased from 1.9 to 2.8 V with the increasing concentration from 1 to 10 M [41]. At different concentrations, the electrochemical stability windows of the aqueous sodium-ion electrolytes based on different sodium salts are shown in Figure 5.3. A highly concentrated $NaClO_4$ of 17 M has an expanded electrochemical stability window of 2.7 V.

Ionic conductivity is a significant property of aqueous electrolytes that will influence the performance of aqueous sodium-ion batteries. Compared with organic electrolytes, aqueous electrolytes usually have high dissociation constant, solubility, and lower viscosity, which endow

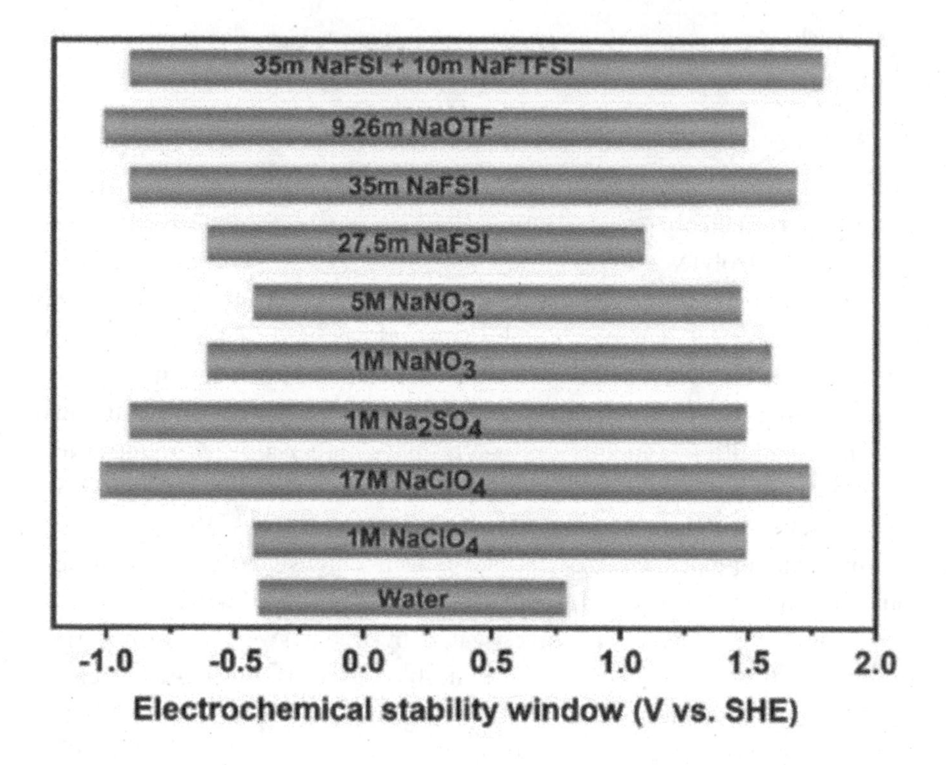

FIGURE 5.3 Electrochemical stability windows of the aqueous sodium-ion electrolytes based on different sodium salts at different concentrations. (Reproduced from Ref. [7] with permission of the American Chemical Society.)

them with higher ionic conductivity than nonaqueous electrolytes. The ionic conductivities of aqueous electrolytes largely depend on the concentrations of inorganic salts, which increase linearly with the electrolyte concentration in a specific limit and then show some nonlinear behavior after that. Under high concentrations, the increased viscosity and the proximity of charged ions in the electrolyte suppress the movement of ions due to the increased electrostatic interactions between different ions in the electrolyte [42].

5.1.4 Summary

Benefited from the low cost, ensured safety, sustainability, and environmental benignity, aqueous sodium-ion batteries are promising alternatives for large-scale energy storage systems. However, the energy density of current aqueous sodium-ion batteries needs to be further increased for better large-scale energy storage applications. The cycle stability also needs to be improved to meet long-term (more than 10 years) energy storage requirements. Therefore, both the scientific and technical challenges of aqueous sodium-ion batteries should be handled in future research for wide and practical energy storage applications on a large scale.

Flexible and wearable batteries have received increasing attention with the growing market of wearable electronics. However, traditional flexible lithium-ion batteries suffer from safety issues from flammable and toxic organic electrolytes, leading to fires and explosions in some conditions such as short circuits and overcharging. Particularly, flexible aqueous sodium-ion batteries which use a nontoxic and neutral aqueous solution of sodium ions can fundamentally resolve the abovementioned potential risks, which are promising for wearable and implantable electronic devices with high requirements for safety.

5.2 FIBER AQUEOUS SODIUM-ION BATTERIES

Fiber aqueous sodium-ion batteries with unique one-dimensional architecture have the advantages of superior flexibility, safety, compatibility, and minimally invasive implantation through injection, promising for wearable and implantable applications, such as pressure monitor, ion concentration meter, and electric pill, due to the inherent biocompatibility and safety of neutral sodium-ion aqueous solutions.

A new family of fiber flexible aqueous sodium-ion batteries had been realized from carbon nanotube/$Na_{0.44}MnO_2$ (CNT/NMO) and CNT/-carbon-coated $NaTi_2(PO_4)_3$ (CNT/NTPO@C)-based hybrid fiber electrodes as cathode and anode (Figure 5.4a) and various aqueous sodium-ion solutions as electrolytes [43]. The NMO and NTPO@C nanoparticles were twisted in the aligned CNT sheets to prepare fiber electrodes (Figure 5.4b–g). Both NMO and NTPO@C particles were well wrapped within the aligned CNTs, and no extra conductive agent or metal current collectors was used in the fiber electrodes.

The fiber aqueous sodium-ion battery demonstrated a discharge specific capacity of 46 mAh g^{-1} (based on the mass of NMO cathode material) at the current density of 0.1 A g^{-1} in pre-deoxygenated 1 M Na_2SO_4 (Figure 5.5a). Even at a higher current density of 3 A g^{-1}, the fiber aqueous sodium-ion battery still delivered a discharge capacity of 12 mAh g^{-1}. The fiber aqueous battery also exhibited high cycling stability, with the capacity maintained at 76% even after 100 cycles (Figure 5.5b). The

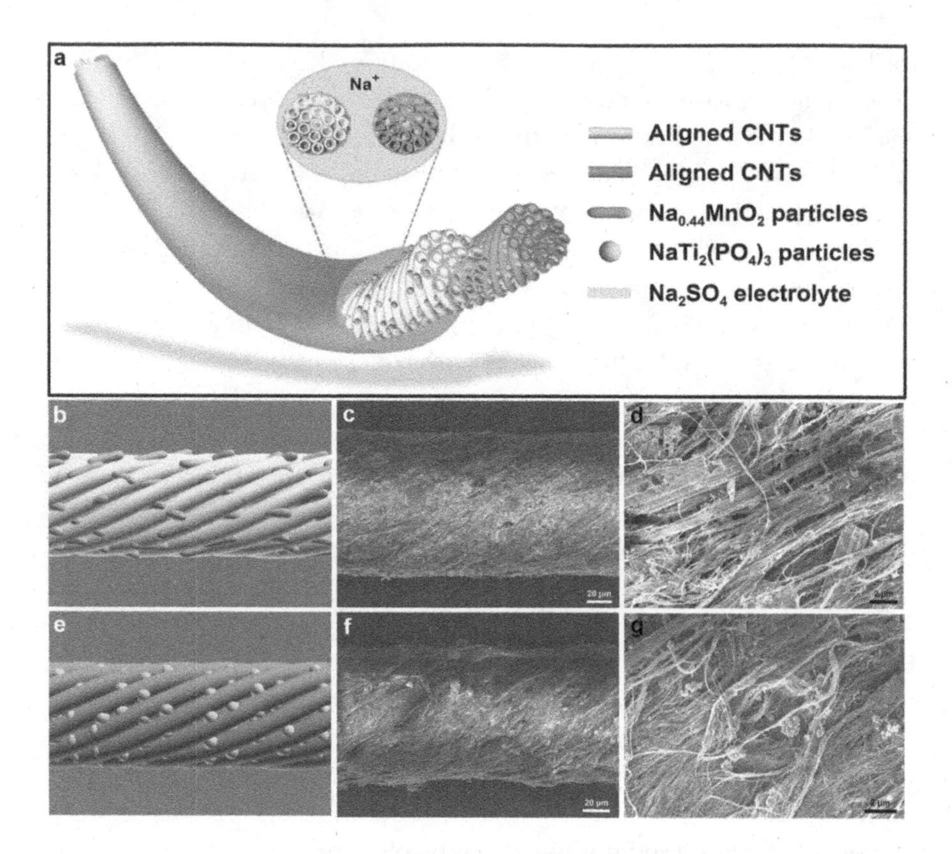

FIGURE 5.4 (a) Schematic diagram of the fiber aqueous sodium-ion battery. (b) Schematic diagram of the carbon nanotube (CNT)/$Na_{0.44}MnO_2$ hybrid electrode. (c and d) SEM images of the CNT/$Na_{0.44}MnO_2$ hybrid electrode at low and high magnifications, respectively. (e) Schematic illustration of the CNT/$NaTi_2(PO_4)_3$@C hybrid electrode. (f and g) SEM images of the CNT/$NaTi_2(PO_4)_3$@C hybrid electrode at low and high magnifications, respectively. (Reproduced from Ref. [43] with permission of Elsevier.)

fiber battery delivered an energy density of 25.7 mWh cm^{-3} at a specific power density of 0.054 W cm^{-3} and an energy density of 5.9 mWh cm^{-3} at a high power density of 0.7 W cm^{-3} (Figure 5.5c). The galvanostatic charge-discharge curves of the battery remained almost unchanged after bending at 180° for 100 times, showing high flexibility (Figure 5.5d). In particular, the fiber aqueous sodium-ion battery could typically work in pre-deoxygenated normal saline (0.9 wt.% NaCl) and cell culture medium (Dulbecco's modified Eagle's medium (DMEM)), showing high safety (Figure 5.5e and f).

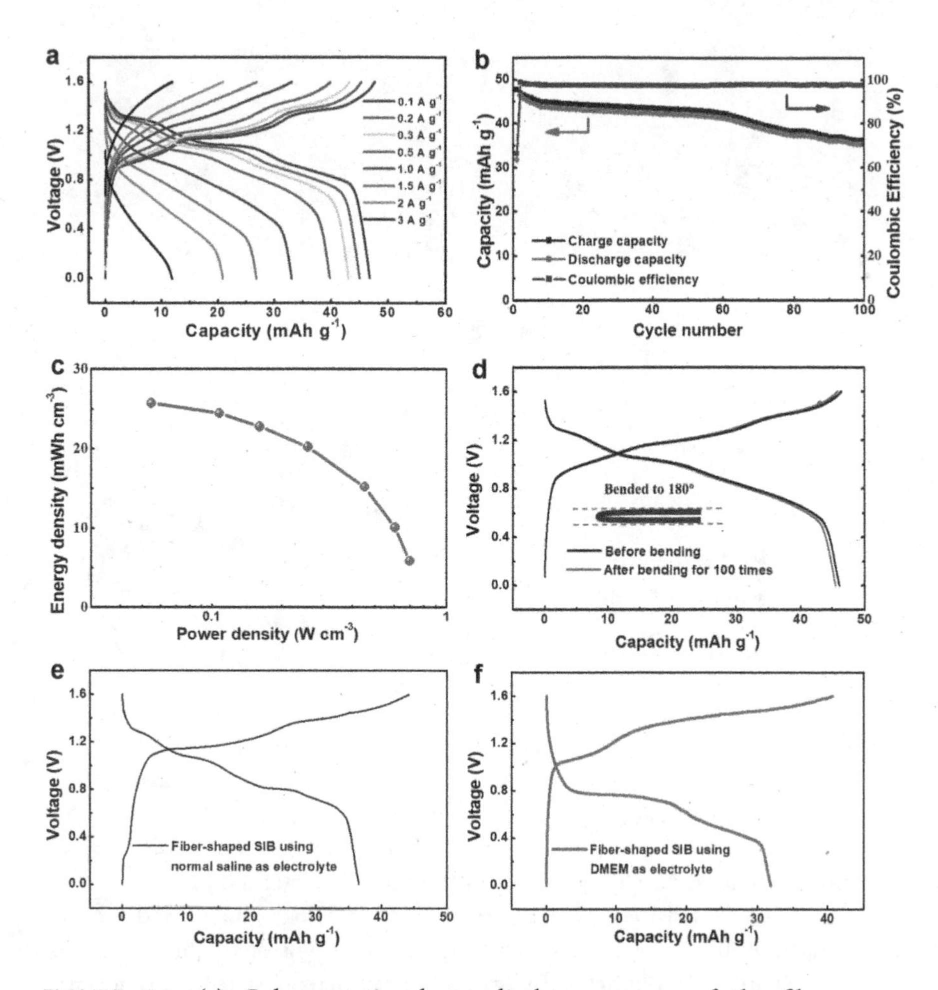

FIGURE 5.5 (a) Galvanostatic charge-discharge curves of the fiber aqueous sodium-ion battery at different current densities. (b) Cycling performance of the fiber aqueous sodium-ion battery. (c) Ragone plot of the fiber aqueous sodium-ion batteries (volumetric energy and power densities). (d) Galvanostatic charge-discharge curves of the fiber battery before and after bending at the current density of 0.2 A g^{-1}. (e and f) Galvanostatic charge-discharge curves of the aqueous sodium-ion batteries using the normal saline and Dulbecco's modified Eagle's medium (DMEM) as electrolytes at the current density of 0.2 A g^{-1}. (Reproduced from Ref. [43] with permission of Elsevier.)

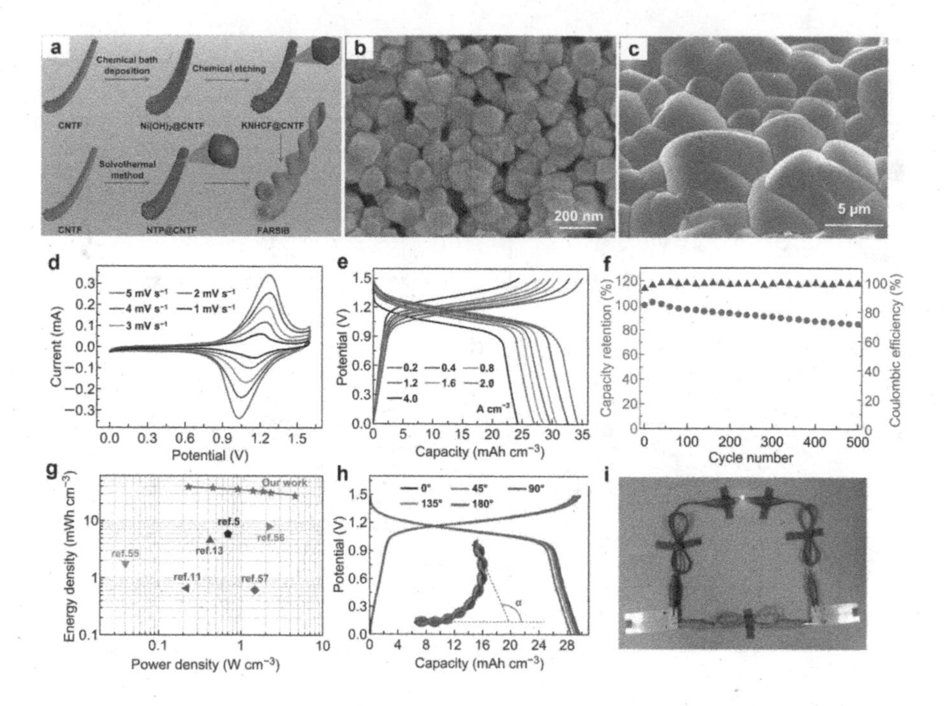

FIGURE 5.6 (a) Schematic diagram of the fabrication process of the quasi-solid-state fiber aqueous sodium-ion battery. (b) SEM image of the KNHCF@CNTF cathode. (c) SEM image of the NTP@CNTF anode. (d) CV curves of the fiber aqueous sodium-ion battery at different scan rates. (e) Galvanostatic charge-discharge curves of the fiber aqueous sodium-ion battery at different current densities. (f) Cycling stability of the fiber aqueous sodium-ion battery during 500 cycles at 2.0 A cm^{-3}. (g) Ragone plot of the fiber aqueous sodium-ion batteries (volumetric energy and power densities). (h) Galvanostatic charge-discharge curves of the fiber battery under different bending angles at 1.2 A cm^{-3}. (i) Application demonstration of the fiber aqueous sodium-ion batteries. (Reproduced from Ref. [44] with permission of Springer.)

Quasi-solid-state fiber aqueous sodium-ion batteries had been further achieved [44] (Figure 5.6a) by using the flexible binder-free $KNiFe(CN)_6$@ carbon nanotube fibers (KNHCF@CNTF) cathode (Figure 5.6b) and $NaTi_2(PO_4)_3$@CNTF (NTP@CNTF) anode (Figure 5.6c). The CV curves of fiber aqueous sodium-ion battery showed a pair of redox peaks with the scan rates increasing from 1 to 5 mV s^{-1} (Figure 5.6d). The fiber aqueous battery delivered a volumetric capacity of 34.21 mAh cm^{-3} at 0.2 A cm^{-3} and a capacity of 24.23 mAh cm^{-3} at 4.0 A cm^{-3} (Figure 5.6e), showing good rate performance. The aqueous battery also showed good cycling stability with 84.7% of the capacity maintained after 500 cycles at 2.0 A cm^{-3} (Figure 5.6f).

The fiber aqueous sodium-ion batteries demonstrated a maximum volumetric energy density of 39.32 mWh cm^{-3} and a maximum volumetric power density of 4.60 W cm^{-3} (Figure 5.6g). They also showed high mechanical flexibility (Figure 5.6h). The galvanostatic charge-discharge curves showed almost negligible changes under different bending angles at 1.2 A cm^{-3}. Two fiber aqueous sodium-ion batteries connected in series could power an LED bulb (Figure 5.6i).

In summary, fiber aqueous sodium-ion batteries based on nontoxic aqueous electrolytes demonstrate high safety, good flexibility, and rate capability, which are promising as safe power sources for wearable electrical devices. Fiber aqueous sodium-ion batteries that use normal biocompatible saline and cell culture medium as electrolytes also have an application prospect in implantable electronic devices.

5.3 IMPLANTABLE FIBER AQUEOUS SODIUM-ION BATTERIES

Implantable electronic devices have been developed rapidly in the recent decade, which can realize physiological monitoring and modulation [45,46], disease treatment, and artificial enhancement [47]. Therefore, it is critical to discover matching power systems specifically for these implantable electronic devices [48,49]. However, most of the existing implanted batteries are bulky and rigid with high Young's moduli (Figure 5.7), which are incompatible with mechanically soft and dynamic biological tissues and may lead to irritations and foreign body immune responses [50]. Therefore, it is critical to developing new kinds of implantable batteries with high safety, softness, and reliability to address the energy supply issues for *in vivo* diagnosis and therapy.

A biocompatible, rechargeable, and implantable fiber aqueous sodium-ion battery had been realized and can be injected into the body for power supply *in vivo* (Figure 5.8a) [54]. The implantable fiber battery used a twisting structure with aligned CNT/NMO as hybrid fiber cathode and CNT/molybdenum trioxide/polypyrrole (CNT/MoO_3/PPy) as hybrid fiber anode (Figure 5.8b). The hybrid fiber electrodes could be knotted without breaking, showing flexibility and robustness (Figure 5.8c). The internal stress of the fiber electrode was proven to be much smaller than conventional conductive fibers such as metal wires, which could provide the fiber

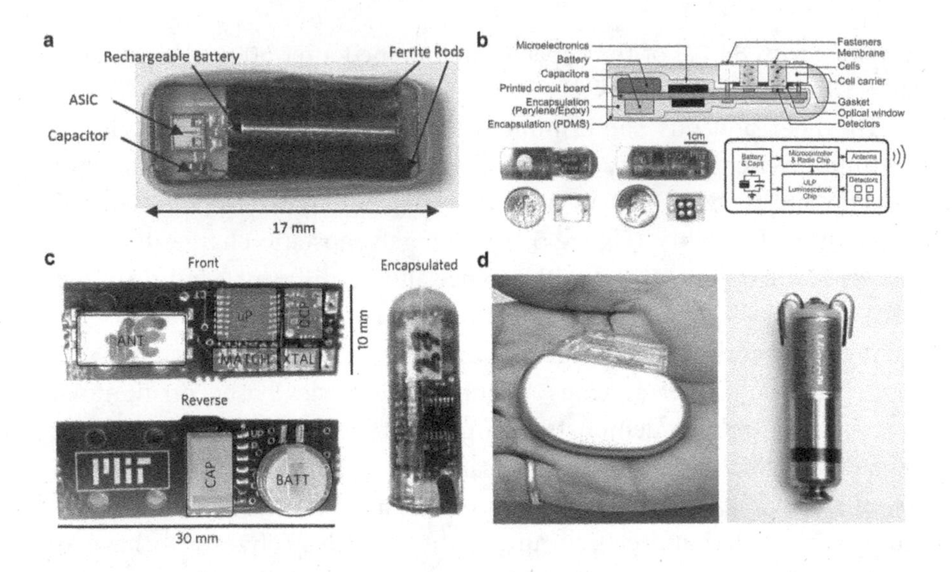

FIGURE 5.7 (a) Photograph of the implanted wirelessly charging battery system for bladder pressure monitoring [51]. (b) The button battery inside the capsule as the power source for gastrointestinal health monitoring [52]. (c) Energy-harvesting primary battery for continuous temperature sensing *in vivo* [53]. (d) Photographs of the commercial cardiac pacemaker (left) and its battery (right). (Reproduced from Ref. [54] with permission of the Royal Society of Chemistry.)

electrode with high mechanical compatibility with tissues (Figure 5.8d). Furthermore, no immune response was found in representative hematoxylin and eosin (H&E) stainings and fluorescence signals of the subcutaneous tissues implanted with the fiber battery after 30 days, demonstrating excellent biocompatibility of the fiber battery (Figure 5.8e–g).

Due to the high flexibility and small size of the fiber aqueous sodium-ion battery, the unencapsulated battery can be injected into the body by a syringe with the assistance of biofluid (Figure 5.9a and b). Notably, the implantable fiber batteries were demonstrated to be implanted into three specific regions: subcutis, heart, and brain of a live mouse through injection process (Figure 5.9c–e). The implanted fiber batteries could closely contact the target tissues and demonstrated specific discharge capacities of 43.05, 39.36, and 33.53 mAh g^{-1} in the mouse's subcutis, heart, and brain, respectively (Figure 5.9f–h). As an application demonstration, the fiber battery was injected into the subcutis of a mouse (Figure 5.9i) to efficiently power an implanted sensor for *in vivo* respiration monitoring

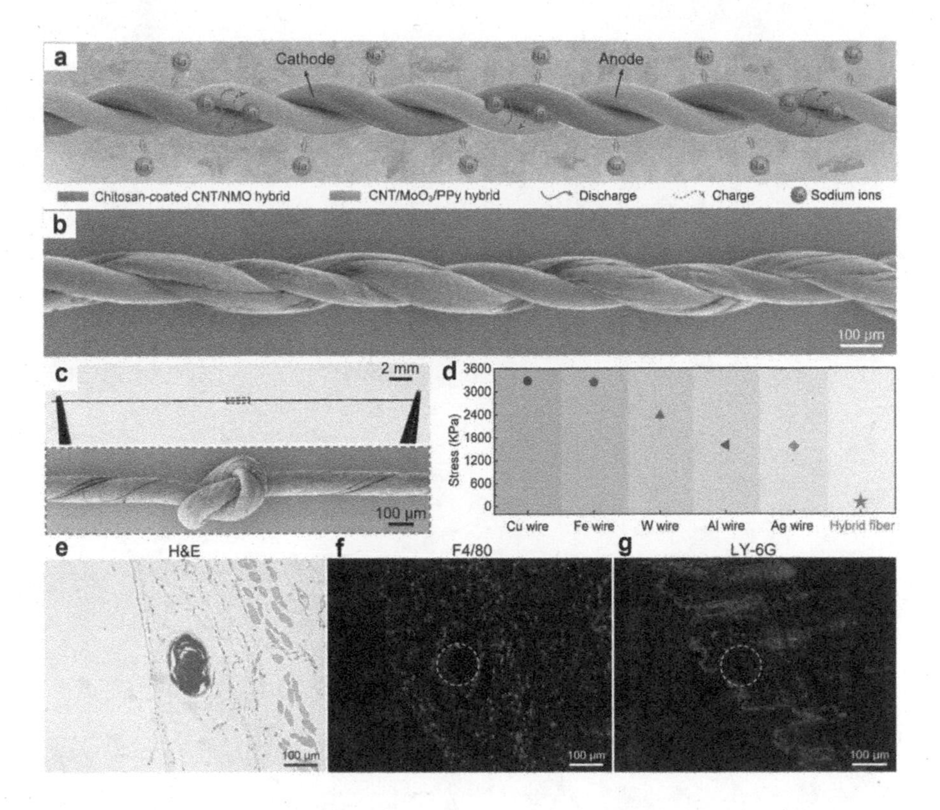

FIGURE 5.8 (a) Working mechanism of the charge/discharge processes of implantable fiber aqueous sodium-ion battery in body fluids. (b) SEM image of the implantable fiber battery with a twisting structure. (c) Photograph and SEM image of the knotted hybrid fiber electrode. (d) The internal stresses of the hybrid fiber electrode and other conductive wires under compression. (e–g) Representative H&E stainings, F4/80-labeled, and LY-6G-labeled sections of subcutaneous tissues with implanted fiber battery after 30 days, respectively. (Reproduced from Ref. [54] with permission of the Royal Society of Chemistry.)

(Figure 5.9j). The implantable fiber batteries offer many advantages over existing implanted batteries and provide new opportunities to design new kinds of implantable and flexible batteries.

5.4 PERSPECTIVE

Traditional lithium/sodium-ion batteries are based on toxic and flammable organic electrolytes, which might cause safety hazards such as fires and explosions in some working conditions such as short circuits and overcharging. Compared with traditional batteries, aqueous sodium-ion

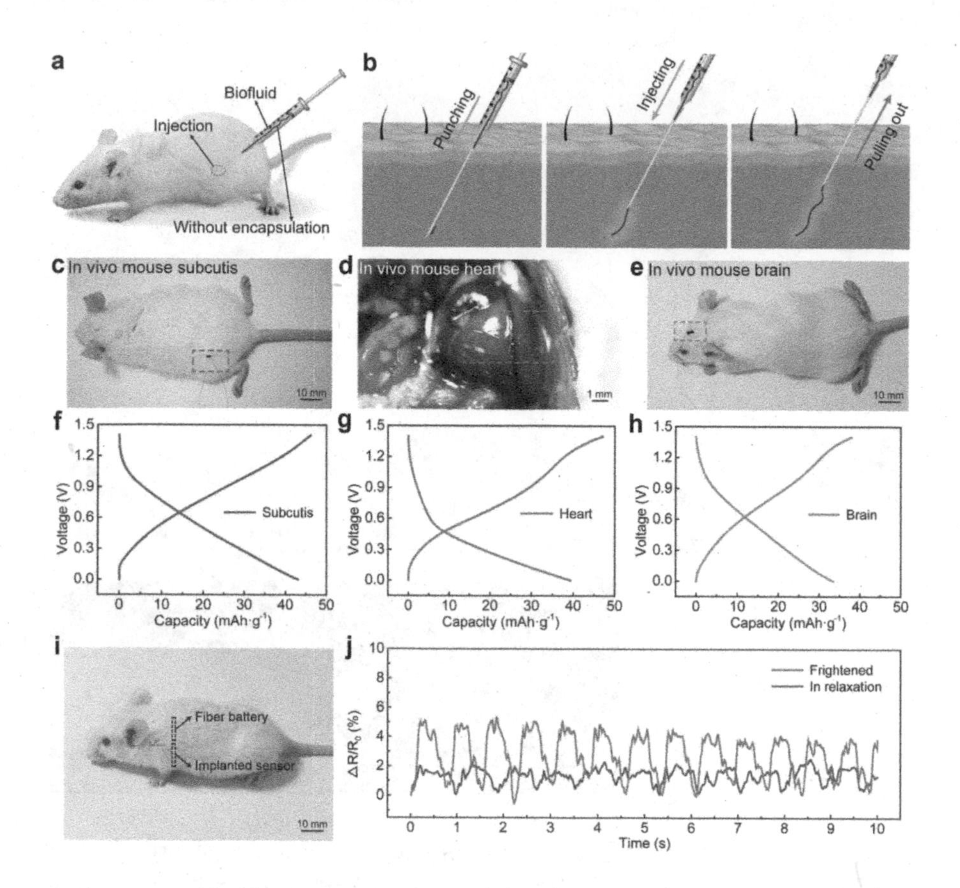

FIGURE 5.9 (a) Schematic diagram of the implantable fiber aqueous sodium-ion battery injected into the body through a syringe. (b) Schematic illustration of the injection process of the fiber battery. (c–e) Photographs of the fiber batteries being injected into the subcutis, heart, and brain of a live mouse, respectively. (f–h) Galvanostatic charge-discharge curve of the implantable fiber batteries in the subcutis, heart, and brain of a live mouse at the current density of 1,000 mA g^{-1}, respectively. (i) Photograph of the mouse implanted with fiber battery and sensor. (j) Respiration monitoring signals recorded from the implanted sensor powering by the implanted fiber battery. (Reproduced from Ref. [54] with permission of the Royal Society of Chemistry.)

batteries use aqueous electrolytes based on sodium salts (such as Na_2SO_4 and NaCl) and possess advantages of ensured safety, fast ion transportation, and lower manufacturing cost. Particularly, one-dimensional fiber aqueous sodium-ion batteries have the advantages of superior flexibility, safety, compatibility, and minimally invasive implantation through injection, which are promising for wearable and implantable electronics.

Despite the progress of fiber aqueous sodium-ion batteries, some challenges still exist and hinder their applications in various fields. On the one hand, the energy densities of the fiber aqueous sodium-ion batteries are still lower than those of conventional fiber batteries due to the relatively narrow electrochemical stability windows of aqueous electrolytes, which need to be improved through different strategies. On the other hand, to ensure the stability of the electrochemical performance during long-term use, it is necessary to develop effective packaging materials and technologies for fiber aqueous sodium-ion batteries.

REFERENCES

1. Luo, J. Y. Xia, Y. Y. 2007. Aqueous lithium-ion battery $LiTi_2(PO_4)_3/LiMn_2O_4$ with high power and energy densities as well as superior cycling stability. *Advanced Functional Materials* 17: 3877–3884.
2. Kim, S. W. Seo, D. H. Ma, X. Ceder, G. Kang, K. 2012. Electrode materials for rechargeable sodium-ion batteries: potential alternatives to current lithium-ion batteries. *Advanced Energy Materials* 2: 710–721.
3. Guo, S. Yi, J. Sun, Y. Zhou, H. 2016. Recent advances in titanium-based electrode materials for stationary sodium-ion batteries. *Energy & Environmental Science* 9: 2978–3006.
4. Yabuuchi, N. Kubota, K. Dahbi, M. Komaba, S. 2014. Research development on sodium-ion batteries. *Chemical Reviews* 114: 11636–11682.
5. Slater, M. Kim, D. Lee, E. Johnson, C. S. 2013. Sodium-ion batteries. *Advanced Functional Materials* 23: 947–958.
6. Hwang, J.-Y. Myung, S.-T. Sun, Y.-K. 2017. Sodium-ion batteries: present and future. *Chemical Society Reviews* 46: 3529–3614.
7. Pahari, D. Puravankara, S. 2020. Greener, safer, and sustainable batteries: an insight into aqueous electrolytes for sodium-ion batteries. *ACS Sustainable Chemistry & Engineering* 8: 10613–10625.
8. Suo, L. Borodin, O. Wang, Y. Rong, X. Sun, W. Fan, X. Xu, S. Schroeder, M. A. Cresce, A. V. Wang, F. 2017. "Water-in-salt" electrolyte makes aqueous sodium-ion battery safe, green, and long-lasting. *Advanced Energy Materials* 7: 1701189.
9. Suo, L. Borodin, O. Gao, T. Olguin, M. Ho, J. Fan, X. Luo, C. Wang, C. Xu, K. 2015. "Water-in-salt" electrolyte enables high-voltage aqueous lithium-ion chemistries. *Science* 350: 938–943.
10. Bin, D. Wang, F. Tamirat, A. G. Suo, L. Wang, Y. Wang, C. Xia, Y. 2018. Progress in aqueous rechargeable sodium-ion batteries. *Advanced Energy Materials* 8: 1703008.
11. Luo, J.-Y. Cui, W.-J. He, P. Xia, Y.-Y. 2010. Raising the cycling stability of aqueous lithium-ion batteries by eliminating oxygen in the electrolyte. *Nature Chemistry* 2: 760–765.

12. Li, Z. Young, D. Xiang, K. Carter, W. C. Chiang, Y. M. 2013. Towards high power high energy aqueous sodium-ion batteries: the $NaTi_2(PO_4)_3$/$Na_{0.44}MnO_2$ system. *Advanced Energy Materials* 3: 290–294.
13. Komaba, S. Ogata, A. Tsuchikawa, T. 2008. Enhanced supercapacitive behaviors of birnessite. *Electrochemistry Communications* 10: 1435–1437.
14. Whitacre, J. Wiley, T. Shanbhag, S. Wenzhuo, Y. Mohamed, A. Chun, S. Weber, E. Blackwood, D. Lynch-Bell, E. Gulakowski, J. 2012. An aqueous electrolyte, sodium ion functional, large format energy storage device for stationary applications. *Journal of Power Sources* 213: 255–264.
15. Jung, Y. H. Lim, C. H. Kim, J.-H. Kim, D. K. 2014. $Na_2FeP_2O_7$ as a positive electrode material for rechargeable aqueous sodium-ion batteries. *RSC Advances* 4: 9799–9802.
16. Zhang, Q. Liao, C. Zhai, T. Li, H. 2016. A high rate 1.2 V aqueous sodium-ion battery based on all NASICON structured $NaTi_2(PO_4)_3$ and $Na_3V_2(PO_4)_3$. *Electrochimica Acta* 196: 470–478.
17. Qin, H. Song, Z. Zhan, H. Zhou, Y. 2014. Aqueous rechargeable alkali-ion batteries with polyimide anode. *Journal of Power Sources* 249: 367–372.
18. Pasta, M. Wessells, C. D. Liu, N. Nelson, J. McDowell, M. T. Huggins, R. A. Toney, M. F. Cui, Y. 2014. Full open-framework batteries for stationary energy storage. *Nature Communications* 5: 1–9.
19. Koshika, K. Sano, N. Oyaizu, K. Nishide, H. 2009. An ultrafast chargeable polymer electrode based on the combination of nitroxide radical and aqueous electrolyte. *Chemical Communications* 836–838.
20. Wang, Y. Mu, L. Liu, J. Yang, Z. Yu, X. Gu, L. Hu, Y. S. Li, H. Yang, X. Q. Chen, L. 2015. A novel high capacity positive electrode material with tunnel-type structure for aqueous sodium-ion batteries. *Advanced Energy Materials* 5: 1501005.
21. Wang, H. Zhang, T. Chen, C. Ling, M. Lin, Z. Zhang, S. Pan, F. Liang, C. 2018. High-performance aqueous symmetric sodium-ion battery using NASICON-structured $Na_2VTi(PO_4)_3$. *Nano Research* 11: 490–498.
22. Zhang, F. Li, W. Xiang, X. Sun, M. 2017. Nanocrystal-assembled porous $Na_3MgTi(PO_4)_3$ aggregates as highly stable anode for aqueous sodium-ion batteries. *Chemistry-A European Journal* 23: 12944–12948.
23. Qu, Q. Liu, L. Wu, Y. Holze, R. 2013. Electrochemical behavior of $V_2O_5{\cdot}0.6H_2O$ nanoribbons in neutral aqueous electrolyte solution. *Electrochimica Acta* 96: 8–12.
24. Deng, C. Zhang, S. Dong, Z. Shang, Y. 2014. 1D nanostructured sodium vanadium oxide as a novel anode material for aqueous sodium ion batteries. *Nano Energy* 4: 49–55.
25. Ke, L. Dong, J. Lin, B. Yu, T. Wang, H. Zhang, S. Deng, C. 2017. A $NaV_3(PO_4)_3$@C hierarchical nanofiber in high alignment: exploring a novel high-performance anode for aqueous rechargeable sodium batteries. *Nanoscale* 9: 4183–4190.

26. Liu, Y. Zhang, B. Xiao, S. Liu, L. Wen, Z. Wu, Y. 2014. A nanocomposite of MoO_3 coated with PPy as an anode material for aqueous sodium rechargeable batteries with excellent electrochemical performance. *Electrochimica Acta* 116: 512–517.
27. Choi, W. Harada, D. Oyaizu, K. Nishide, H. 2011. Aqueous electrochemistry of poly(vinylanthraquinone) for anode-active materials in high-density and rechargeable polymer/air batteries. *Journal of the American Chemical Society* 133: 19839–19843.
28. Fernández-Ropero, A. Saurel, D. Acebedo, B. Rojo, T. Casas-Cabanas, M. 2015. Electrochemical characterization of $NaFePO_4$ as positive electrode in aqueous sodium-ion batteries. *Journal of Power Sources* 291: 40–45.
29. Chang, Z. Yang, Y. Li, M. Wang, X. Wu, Y. 2014. Green energy storage chemistries based on neutral aqueous electrolytes. *Journal of Materials Chemistry A* 2: 10739–10755.
30. Wang, Y. Yi, J. Xia, Y. 2012. Recent progress in aqueous lithium-ion batteries. *Advanced Energy Materials* 2: 830–840.
31. Wu, W. Shabhag, S. Chang, J. Rutt, A. Whitacre, J. F. 2015. Relating electrolyte concentration to performance and stability for $NaTi_2(PO_4)_3$/$Na_{0.44}MnO_2$ aqueous sodium-ion batteries. *Journal of The Electrochemical Society* 162: A803.
32. Wang, Y. Liu, J. Lee, B. Qiao, R. Yang, Z. Xu, S. Yu, X. Gu, L. Hu, Y.-S. Yang, W. 2015. Ti-substituted tunnel-type $Na_{0.44}MnO_2$ oxide as a negative electrode for aqueous sodium-ion batteries. *Nature Communications* 6: 1–10.
33. Luo, D. Lei, P. Huang, Y. Tian, G. Xiang, X. 2019. Improved electrochemical performance of graphene-integrated $NaTi_2(PO_4)_3$/C anode in high-concentration electrolyte for aqueous sodium-ion batteries. *Journal of Electroanalytical Chemistry* 838: 66–72.
34. Nian, Q. Liu, S. Liu, J. Zhang, Q. Shi, J. Liu, C. Wang, R. Tao, Z. Chen, J. 2019. All-climate aqueous dual-ion hybrid battery with ultrahigh rate and ultralong life performance. *ACS Applied Energy Materials* 2: 4370–4378.
35. Reddy, R. N. Reddy, R. G. 2003. Sol-gel MnO_2 as an electrode material for electrochemical capacitors. *Journal of Power Sources* 124: 330–337.
36. Kim, D. J. Ponraj, R. Kannan, A. G. Lee, H.-W. Fathi, R. Ruffo, R. Mari, C. M. Kim, D. K. 2013. Diffusion behavior of sodium ions in $Na_{0.44}MnO_2$ in aqueous and non-aqueous electrolytes. *Journal of Power Sources* 244: 758–763.
37. Li, Z. Ravnsbæk, D. B. Xiang, K. Chiang, Y.-M. 2014. $Na_3Ti_2(PO_4)_3$ as a sodium-bearing anode for rechargeable aqueous sodium-ion batteries. *Electrochemistry Communications* 44: 12–15.
38. Chen, L. Liu, J. Guo, Z. Wang, Y. Wang, C. Xia, Y. 2016. Electrochemical profile of $LiTi_2(PO_4)_3$ and $NaTi_2(PO_4)_3$ in lithium, sodium or mixed ion aqueous solutions. *Journal of the Electrochemical Society* 163: A904.
39. Song, W. Ji, X. Zhu, Y. Zhu, H. Li, F. Chen, J. Lu, F. Yao, Y. Banks, C. E. 2014. Aqueous sodium-ion battery using a $Na_3V_2(PO_4)_3$ electrode. *ChemElectroChem* 1: 871–876.

40. Gao, H. Goodenough, J. B. 2016. An aqueous symmetric sodium-ion battery with NASICON-structured $Na_3MnTi(PO_4)_3$. *Angewandte Chemie International Edition* 128: 12960–12964.
41. Nakamoto, K. Sakamoto, R. Ito, M. Kitajou, A. Okada, S. 2017. Effect of concentrated electrolyte on aqueous sodium-ion battery with sodium manganese hexacyanoferrate cathode. *Electrochemistry* 85: 179–185.
42. Hamann, C. Hamnett, A. Vielstich, W. 2007. *Electrochemistry*, 2nd completely revised and updated ed. Weinheim: Wiley-VCH Pub 98–102.
43. Guo, Z. Zhao, Y. Ding, Y. Dong, X. Chen, L. Cao, J. Wang, C. Xia, Y. Peng, H. Wang, Y. 2017. Multi-functional flexible aqueous sodium-ion batteries with high safety. *Chem* 3: 348–362.
44. He, B. Man, P. Zhang, Q. Fu, H. Zhou, Z. Li, C. Li, Q. Wei, L. Yao, Y. 2019. All binder-free electrodes for high-performance wearable aqueous rechargeable sodium-ion batteries. *Nano-Micro Letters* 11: 1–12.
45. Kang, S.-K. Murphy, R. K. Hwang, S.-W. Lee, S. M. Harburg, D. V. Krueger, N. A. Shin, J. Gamble, P. Cheng, H. Yu, S. 2016. Bioresorbable silicon electronic sensors for the brain. *Nature* 530: 71–76.
46. Kim, T.-I. McCall, J. G. Jung, Y. H. Huang, X. Siuda, E. R. Li, Y. Song, J. Song, Y. M. Pao, H. A. Kim, R.-H. 2013. Injectable, cellular-scale optoelectronics with applications for wireless optogenetics. *Science* 340: 211–216.
47. Canales, A. Jia, X. Froriep, U. P. Koppes, R. A. Tringides, C. M. Selvidge, J. Lu, C. Hou, C. Wei, L. Fink, Y. 2015. Multifunctional fibers for simultaneous optical, electrical and chemical interrogation of neural circuits in vivo. *Nature Biotechnology* 33: 277–284.
48. Chae, J. S. Heo, N.-S. Kwak, C. H. Cho, W.-S. Seol, G. H. Yoon, W.-S. Kim, H.-K. Fray, D. J. Vilian, A. E. Han, Y.-K. 2017. A biocompatible implant electrode capable of operating in body fluids for energy storage devices. *Nano Energy* 34: 86–92.
49. Huang, S. Liu, Y. Zhao, Y. Ren, Z. Guo, C. F. 2019. Flexible electronics: stretchable electrodes and their future. *Advanced Functional Materials* 29: 1805924.
50. Park, S. Heo, S. W. Lee, W. Inoue, D. Jiang, Z. Yu, K. Jinno, H. Hashizume, D. Sekino, M. Yokota, T. 2018. Self-powered ultra-flexible electronics via nano-grating-patterned organic photovoltaics. *Nature* 561: 516–521.
51. Young, D. J. Cong, P. Suster, M. A. Damaser, M. 2015. Implantable wireless battery recharging system for bladder pressure chronic monitoring. *Lab on a Chip* 15: 4338–4347.
52. Mimee, M. Nadeau, P. Hayward, A. Carim, S. Flanagan, S. Jerger, L. Collins, J. McDonnell, S. Swartwout, R. Citorik, R. J. 2018. An ingestible bacterial-electronic system to monitor gastrointestinal health. *Science* 360: 915–918.

53. Nadeau, P. El-Damak, D. Glettig, D. Kong, Y. L. Mo, S. Cleveland, C. Booth, L. Roxhed, N. Langer, R. Chandrakasan, A. P. 2017. Prolonged energy harvesting for ingestible devices. *Nature Biomedical Engineering* 1: 1–8.
54. Zhao, Y. Mei, T. Ye, L. Li, Y. Wang, L. Zhang, Y. Chen, P. Sun, X. Wang, C. Peng, H. 2021. Injectable fiber batteries for all-region power supply in vivo. *Journal of Materials Chemistry A* 9: 1463–1470.

CHAPTER 6

Flexible Aqueous Zinc-Ion Batteries

6.1 OVERVIEW OF AQUEOUS ZINC-ION BATTERIES

At present, human society is developing rapidly. The consumption of resources increases year by year, not conducive to future scientific and technological progress and leads to significant environmental damage. Therefore, it is crucial to promote the transition from fossil fuels to renewable energy during the progress. As a critical component to the transition process, energy storage devices are increasingly concerned from academia, making significant progress. Among these energy storage devices, the lithium-ion battery has become the most widely used, benefiting from its high energy density and long cycle life. However, serious problems such as inadequate safety, high costs, and severe environmental pollution have been exposed with the deepening development. On such a basis, developing an energy storage device more in line with the current development is necessary. Aqueous zinc-ion batteries have become a popular candidate for the next generation of energy storage devices due to their high electrochemical performance, ensured safety, low expense, resource abundance, and environmental friendliness. In this section, the material design from anode, cathode, and electrolyte perspectives are primarily focused on. The working mechanisms are concluded based on the introduction of applied materials. The section serves as the fundamentals for further introduction to flexible aqueous zinc-ion batteries.

DOI: 10.1201/9781003273677-6

6.1.1 Materials

Traditional aqueous zinc-ion batteries are mainly composed of three components: anode, cathode, and electrolyte. These parts are connected to form a closed loop to ensure the stable progress of the battery charging and discharging process. For high and steady power output, the material utilized is crucial. In this part, the anode material is firstly focused, after which the cathode and electrolyte material are discussed.

6.1.1.1 Anodes

Zinc metal is used as the anode material in most aqueous zinc-ion batteries benefiting from its abundant reserves, high theoretical capacity (820 $mAh{\cdot}g^{-1}$), and ensured safety. However, despite such advantages, there are substantial problems, such as dendrite growth, corrosion, and passivation, which are unignorable. Dendrites refer to the small protrusions formed *via* uneven Zn electrodeposition during operation, leading to short circuit. Corrosion and passivation reduce the active mass/area of the anode and impact cyclability. To solve these problems, rather than relying on bare Zn metal solely, introducing other materials is vital. Commonly, such strategies are classified into two categories: surface modifications and structural design.

Surface modification refers to coating the Zn anode surface with suitable materials to reduce corrosion and the generation of by-products. The most straightforward strategy is metal-based coating. Due to its inert nature, copper can isolate moisture, can inhibit corrosion, and is applied in aqueous zinc-ion batteries. The fully symmetrical battery composed of Zn/Cu composite could work stably for 1,500 h at 1 $mA{\cdot}cm^{-2}$ and low depth of discharge [1]. Nevertheless, metallic coating suffers from possible corrosion due to its activity. Due to chemical robustness, oxides are also developed as the protective layer. Specifically, ZrO_2, serving as nucleation sites for reversible stripping/deposition, was responsible for stable cycling for 2,200 h at a 5 $mA{\cdot}cm^{-2}$ [2]. However, due to the disadvantage of poor electron conductivity, high voltage hysteresis is observed. Carbonaceous materials are also favored, thanks to their conductivity. For example, a carbon nanotube (CNT) coating technique is recently presented, which inhibited dendrite growth and side reactions [3]. In addition to the above materials, polymers matrix is regarded as a suitable choice. For example, a highly viscoelastic polyvinyl butyral film was applied to prevent the migration of H_2O to the anodic surface, leading to the 2,200 h cycle time of the symmetric cells at 0.5 mA·cm^{-2} [4].

The essence of the structural design is to increase the surface area of the electrode and the collector, which reduces the local current density and promotes the uniform deposition of Zn^{2+}. It is generally achieved by introducing types of materials as the Zn hosts. Carbonaceous materials are again applied due to their high conductivity. Typically, 3D CNTs are used as a framework for a dendritic-free zinc metal anode. The symmetric cell composed such anode cycled stably for more than 100 h at various current densities [5]. Additionally, porous foams with a high surface area are also designed as viable Zn hosts, which improves coulombic efficiency. Specifically, the porous copper foam was applied, reducing the zinc nucleation energy barrier and optimizing the overall coulombic efficiency and long-term stability [6].

6.1.1.2 Cathodes

Compared with anode materials, the research on cathode materials is more in-depth and extensive. At present, the most commonly used cathode materials can be classified into three categories: Mn-based materials, V-based materials, and Prussian blue analogs. This section summarizes the recent advancement in such cathode materials regarding the obstacles and corresponding techniques.

Mn-based cathode materials refer primarily to Mn oxides, such as MnO_2, MnO, Mn_2O_3, and Mn_3O_4. MnO_2, with the highest oxidation state of Mn, is the most extensively used Mn-based cathodes in aqueous zinc-ion batteries. Based on crystallography, σ-MnO_2, β-MnO_2, γ-MnO_2, δ-MnO_2, ε-MnO_2, λ-MnO_2, R-MnO_2, and T-MnO_2 usually are illustrated (Figure 6.1a) [7].

Due to strong John–Teller effects during Mn (IV)–Mn (III) transition, Mn-based materials suffer from dramatic dissolution and subsequent capacity loss, hindering their practical development. Hence, improvements have been made in recent years to suppress Mn's dissolution to improve the cathode's performance. It is found that the instability results from phase change and water intercalation. Polyaniline (PANI)-intercalated layered MnO_2 was fabricated to suppress the phase change, strengthening the layered structure and facilitating charge storage. Therefore, a high specific capacity of 280 mAh·g^{-1} for 200 cycles was achieved [8]. Apart

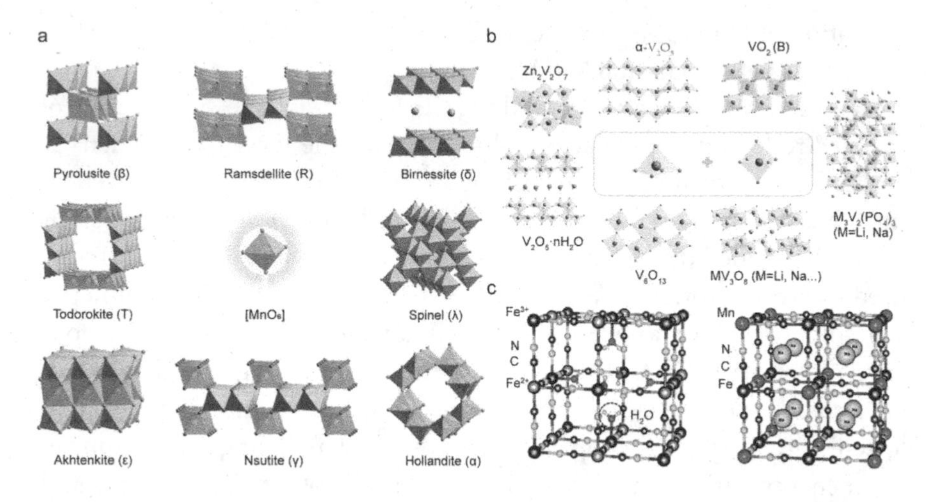

FIGURE 6.1 Schematic illustration of crystallography of (a) Mn-based cathode materials; (b) V-based cathode materials; and (c) Prussian blue analogs. (Reproduced from Ref. [7] with permission of the American Chemical Society.)

from introducing polymer matrix, atomic doping is also an effective strategy. The dopants can be universally chosen from metal and nonmetal elements. Specifically, vanadium could be introduced into MnO_2 through a simple redox reaction, increasing the specific surface area and conductivity, which benefited both discharge voltage and cycle stability [9]. Nitrogen is often used as a doping element due to its unique function in regulating the energy bands. A low-temperature ammonia treatment technique was applied to dope nitrogen into manganese dioxide, improving the MnO_2 conductivity and enabling a stable capacity of 200 $mAh{\cdot}g^{-1}$ for 1,000 cycles at 1 $A{\cdot}g^{-1}$ [10]. Another strategy to regulate the conductivity is to composite MnO_2 with carbonaceous materials. For instance, MnO_2 was deposited on acid-treated CNTs through a redox reaction, and the electrical conductivity was significantly improved, leading to 400 $mAh{\cdot}g^{-1}$ capacity output at 1 $A{\cdot}g^{-1}$ [11].

Apart from MnO_2, some less commonly used Mn-based materials have been investigated, such as MnO. Due to the conventional belief that Mn (II) has weak or no activity, MnO has not been considered a suitable cathode material. However, according to recent research, MnO was gradually converted into layered MnO_2, serving as the Zn^{2+} insertion/extraction reaction site. It was proven that after completing the conversion of layered MnO_2, MnO could reach a high specific capacity of 330 $mAh{\cdot}g^{-1}$ at 0.1 $A{\cdot}g^{-1}$ [12].

As Mn-based materials, V-based materials are also widely applied as cathodes for aqueous zinc-ion batteries. The popularity benefits from their high capacity, abundant reserves, and low cost. At present, the most common V-based materials utilized are vanadium oxides and metal vanadate (Figure 6.1b) [7].

As MnO_2, due to the highest oxidation state, V_2O_5, the most commonly applied one, with its unique layered structure facilitates the insertion and extraction of Zn^{2+}. The surficial morphology evolved from initial smooth to porous with the reversible insertion of Zn^{2+} and H_2O, enabling subsequent zinc storage. The battery composed of such cathode material could reach a capacity of around 400 $mAh{\cdot}g^{-1}$ at 5 $A{\cdot}g^{-1}$ and maintain 99.3% of the initial capacity after 4,000 cycles [13]. H_2O molecules are believed to enlarge the interlayer spacings and thus weaken the electrostatic effect in the V_2O_5 framework and enable fast diffusion kinetics. The as-prepared aqueous zinc-ion batteries exhibited an energy density of 144 $Wh{\cdot}kg^{-1}$ at 0.3 $A{\cdot}g^{-1}$ [14].

Typically, two species of metal vanadates are applied in aqueous zinc-ion batteries, i.e., $M_xV_3O_8$ and $M_xV_2O_7$. $M_xV_3O_8$ is formed by stacking V_3O_8 layer by layer with a stable structure. Specifically, $H_2V_3O_8$ with high performance was developed with an exceptionally durable frame due to the strong hydrogen bonds between H and O atoms. The V's structural stability and valence state conversion prevent undesirable lattice distortion during the Zn^{2+} insertion/extraction [15]. There are other similar vanadates, such as LiV_3O_8 and NaV_3O_8 [16]. Additionally, other vanadates such as $M_xV_2O_7$ are also developed, but the related publications are scarce.

Prussian blue and its derivatives are newly developed as cathode materials for aqueous zinc-ion batteries (Figure 6.1c). Unique advantages have been discovered for Prussian blue analogs. First, the interstitial space inside is extensive, providing enough space to accommodate large amounts of Zn^{2+} without distortion. Second, its spatial structure increases the diversity of atoms combined with it, making it easier to adjust the electrochemical properties. In addition, the multiple redox centers make their theoretical capacity extremely large [17]. However, severe problems within preparation have been observed. H_2O molecules are easily mixed into their internal structure, deforming the entire material and blocking the Zn^{2+} diffusion path. A common way to solving this is to evaporate the crystal water by heat treatment while maintaining a stable framework. The working voltage could thus reach 1.7 V [18].

6.1.1.3 Electrolytes

The most applied electrolytes for aqueous zinc-ion batteries are classified into liquid electrolytes and gel electrolytes. Liquid electrolytes are initially developed for aqueous zinc-ion batteries due to their low cost, high ionic conductivity, and high safety. However, the narrow operation voltage window has been a critical issue to practical application. To this end, other liquid electrolytes such as $Zn(CF_3SO_3)_2$ electrolytes were investigated. By paring Zn anode with $ZnMn_2O_4$, a vast operation voltage window was achieved while maintaining excellent long-cycle performance and a coulombic efficiency close to 100% [19]. However, despite efforts to improve the performance of liquid electrolytes, their intrinsic lack of structural robustness has significantly hindered the application in flexible aqueous zinc-ion batteries. Therefore, gel electrolytes with both high ionic conductivity and self-standing properties are thus developed.

Gel electrolytes belong to solid electrolytes, which generally consist of two forms of all-solid electrolytes and quasi-solid electrolytes, depending on whether H_2O-related substances exist within the framework. All-solid electrolytes are typically composed of polymer matrix and salts. However, their ionic conductivity is limited due to their water-free nature, which is unsuitable for aqueous zinc-ion batteries with high power output. Gel electrolytes combine the advantages of both all-solid electrolytes and liquid electrolytes, ensuring flexibility, safety, and high ionic conductivity. Polyvinyl alcohol (PVA) is a polymer framework primarily used in gel electrolytes, combined with $Zn(CF_3SO_3)_2$ solution to form gel electrolytes for aqueous zinc-ion batteries. The as-fabricated aqueous zinc-ion batteries showed superior electrochemical performance and self-healing capability, delivering an outlook for later development of flexible aqueous zinc-ion batteries for wearable electronic devices [20].

6.1.2 Working Mechanism

Figure 6.2 briefly illustrates the working mechanism of the aqueous zinc-ion batteries with zinc metal as the anode and MnO_2 as the cathode. Zn^{2+} ions are stripped/deposited from the anode in aqueous Zn-based electrolytes and reversibly inserted/extracted from the cathode [21]. The specific reaction can be depicted by the formula below (take MnO_2 as an example):

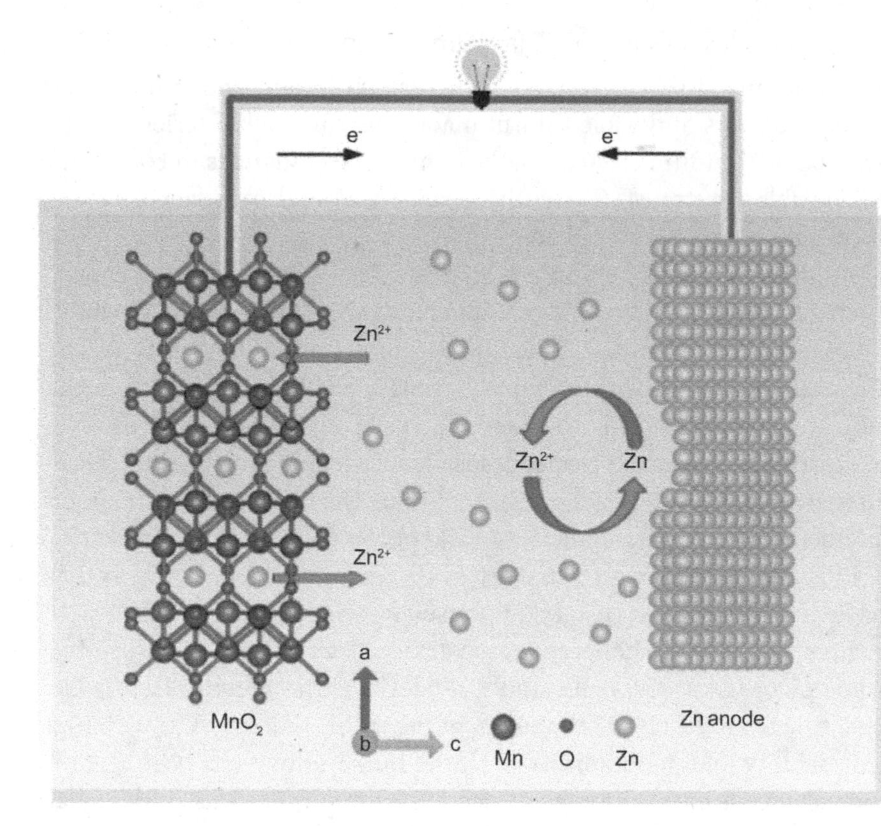

FIGURE 6.2 Schematic illustration of the working mechanism of aqueous zinc-ion batteries. (Reproduced from Ref. [21] with permission of the American Chemical Society.)

Cathode:

$$MnO_2 + xZn^{2+} + 2xe^- \leftrightarrow Zn_xMnO_2$$

Anode:

$$Zn - 2e^- \leftrightarrow Zn^{2+}$$

6.2 FLEXIBLE THIN-FILM AQUEOUS ZINC-ION BATTERIES

The most straightforward construction of flexible thin-film aqueous zinc-ion batteries is the 2D sandwich structure, resembling rigid aqueous zinc-ion batteries in coin and pouch cells. To construct 2D sandwich-like aqueous zinc-ion batteries, key components such as electrodes and electrolytes are thus required to be flexible.

6.2.1 Electrodes for Flexible Thin-Film Aqueous Zinc-Ion Batteries

Traditional metal-based electrodes are generally rigid and cannot meet the requirements of flexible thin-film aqueous zinc-ion batteries. To prepare flexible thin-film electrodes, the common technique is to compound with flexible substrates. Currently, metal-based and carbonaceous substrates have been widely applied.

Thin-film zinc foils exhibit a certain degree of deformability, capable of being bent and twisted within a specific range. However, the flexibility of pure zinc foil is limited. Therefore, to improve the flexibility of zinc metal, compounding with metal materials such as copper foam, nickel foam, and stainless steel is considered a viable strategy [22,23]. However, the deformability of such electrodes is also flawed. Meanwhile, the heavy and inert metal substrate significantly lowers the flexible thin-film aqueous zinc-ion batteries' overall energy and power density. Therefore, carbonaceous materials have attracted attention from academia owing to their excellent flexibility, low cost, and lightweight.

Three primary carbonaceous substrate materials are developed as carbon cloth (CC), graphene, and CNTs. The earliest application is CC. Benefiting from the fast development of the textile industry, CC with high deformability and good affinity with the human body has been used to produce flexible energy storage devices. The flexible aqueous zinc-ion batteries prepared with CC current collectors maintained the identical capacity to the initial performance under nearly 90° bending deformation. Besides, it could be integrated with a particular flexible wearable generator to realize the timely replenishment of energy, providing a theoretical basis for realizing flexible wearable power stations in the future [24]. Graphene is also commonly used for preparing flexible electrodes. Graphene foams with porous structures are particularly applicated. Typically, zinc vanadate arrays designed with graphene foam as a substrate exhibit excellent flexibility and mechanical stability. It can maintain a stable discharge voltage and capacity even after large-angle and repeated bending. At the same time, its unique porous structure promotes the effective transfer of electrons and inhibits the production of dendrites and by-products [25].

Although the flexibility of the above electrode is improved to a certain extent, there is still a distance from the current standards for the flexibility of wearable devices. CNTs, as a novel type of carbonaceous material, are recently developed as electrode substrates due to their excellent

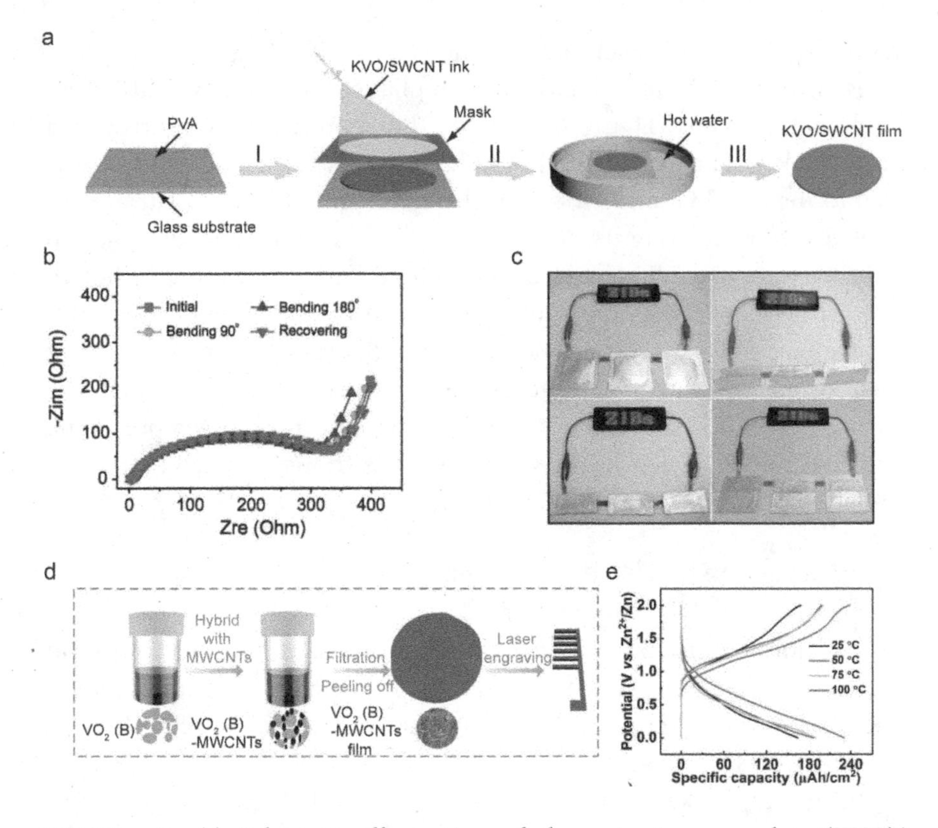

FIGURE 6.3 (a) Schematic illustration of the potassium vanadate (KVO)/single-walled carbon nanotube (SWCNT) film preparation process. (b) Electrochemical impedance spectra before/after bending and after bending recovery. (c) Schematic illustration of soft pack battery application under different deformation conditions. Reproduced from Ref. [26] with permission of the American Chemical Society. (d) Schematic illustration of preparing VO_2-multi-walled carbon nanotubes (MWCNTs) film. (e) Galvanostatic charge-discharge (GCD) curves versus temperature of the flexible aqueous zinc-ion batteries. (Reproduced from Ref. [27] with permission of Wiley-VCH.)

mechanical properties and superior electrochemical, optical, and thermal performance. Potassium vanadate (KVO) cathodes composited with single-walled CNTs (SWCNTs) films (Figure 6.3a) exhibited ultrahigh flexibility, which could restore a stable interface impedance under different bending angles, leading to unchanged cyclic voltammetry (CV) and charge-discharge curves (Figure 6.3b). The specific capacity remained the same as the initial state under various bending angles. Moreover, the as-fabricated soft pack battery showed excellent mechanical properties, which could power electrical appliances stably under different bending conditions (Figure 6.3c) [26]. The application of multi-walled CNTs (MWCNTs) is also highlighted. An electroplating method was used to deposit VO_2 and zinc nanosheets on the MWCNTs surface for flexible electrodes (Figure 6.3d). Flexible thin-film aqueous zinc-ion batteries composed of such electrodes could be bent to 150°. Although the discharge voltage decreased slightly with the increase of the bending angle, it could maintain 97.8% of the initial capacity. In addition, its most unique advantage is high-temperature resistance of up to 100°C (Figure 6.3e). This provides an essential theoretical basis for preparing flexible thin-film aqueous zinc-ion batteries with high capacity and coulombic efficiency [27].

6.2.2 Electrolyte for Flexible Thin-Film Aqueous Zinc-Ion Batteries

This section summarizes the advantages and disadvantages of all-solid electrolytes and gel electrolytes used in flexible thin-film aqueous zinc-ion batteries. The significant advantage of all-solid electrolytes is that it depletes water solvents in traditional electrolytes, eliminating the risk of battery leakage and improving safety and stability. One typical all-solid aqueous zinc-ion battery was made by combining polyvinylidene fluoride hexafluoropropylene (PVDF-HFP) and $Zn(BF_4)_2$ as electrolytes, which could not only suppress the hydrogen evolution reaction and dendrite growth but due to the certain degree of flexibility of such an all-solid electrolyte, it could also maintain the same capacity under bending deformation. Even after being bent many times, the capacity did not change significantly. Also, a unique advantage of withstanding a relatively harsh working temperature was observed [28]. Other related reports include producing an all-solid electrolyte combining 1,3-dioxolane and $Zn(BF_4)_2$ by *in situ* polymerization. Under large and long bending cycles, the capacity retention rate was close to 100%, and the interface resistance did not change significantly. Besides, a high capacity retention rate was discovered after the full aqueous zinc-ion battery was burned, indicating its anti-combustion capability [29]. However, the drawbacks of all-solid electrolytes are unignorable, whose ionic conductivity is inferior, limiting their further development. Therefore, gel electrolytes are more often used in flexible thin-film aqueous zinc-ion batteries.

Compared with liquid and all-solid electrolytes, gel electrolytes combine the advantages of both. Electrolyte leakage is avoided while ionic conductivity is ensured. Besides, due to the less free water content, the material dissolution, e.g., Mn dissolution for Mn-based materials, is suppressed. Also, the polymer in the electrolyte forms a close interface with the Zn anode for an even electrical field, inhibiting the formation of dendrites. Moreover, gel electrolytes also act as the separator and simplify the assembling process.

The primary choice and widely used polymer matrix is PVA. Because of its excellent ionic conductivity and unique self-healing ability, it is welcomed by academia and becomes one of the best matrixes for gel electrolytes. 2D sandwich-like aqueous zinc-ion batteries using $Zn(CF_3SO_3)_2$ accommodated in a PVA-based gel electrolyte as the electrolyte could keep stable capacity under various deformation conditions (Figure 6.4a). The self-healing ability was confirmed by cutting the as-fabricated aqueous zinc-ion batteries several times and adhering them together. The capacity was well restored afterward (Figure 6.4b) [20]. In addition to PVA, polyacrylamide (PAM) is also commonly used. The PAM-based gel electrolytes exhibit excellent mechanical and waterproof properties. An

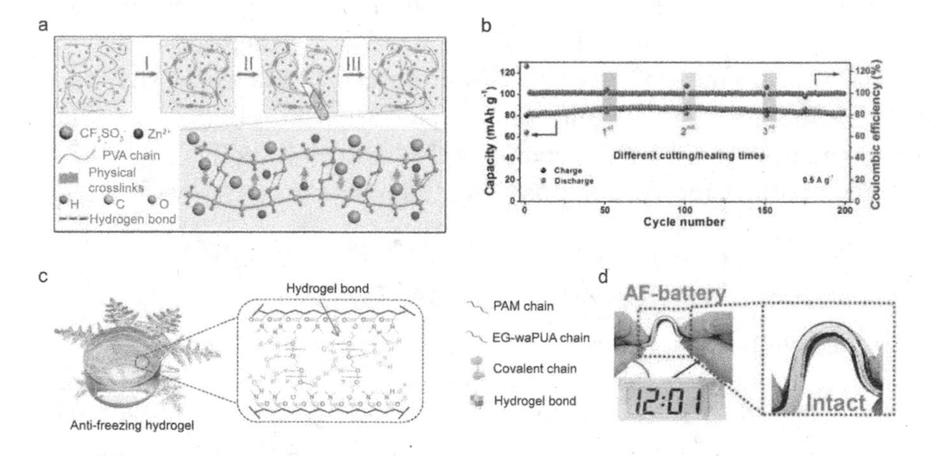

FIGURE 6.4 (a) Schematic process of fabricating self-healing electrolyte and its self-healing illustration. (b) Cycle performance of the self-healing aqueous zinc-ion batteries at the original state and after multiple cutting/self-healing times. Reproduced from Ref. [20] with permission of Wiley-VCH. (c) Schematic illustration of the structure of anti-freezing (AF) gel electrolyte. (d) Demonstration of bending the battery at –20°C. (Reproduced from Ref. [31] with permission of the Royal Society of Chemistry.)

earlier application of PAM is to combine it with gelatin to form a layered polymer electrolyte. Compared with the traditional flexible batteries, the PAM-based aqueous zinc-ion batteries delivered higher safety performance and durability. The waterproof capability was verified by placing the as-prepared aqueous zinc-ion batteries in fire or immersing them in different solutions, where there was no significant effect on its capacity [30]. Furthermore, PAM is also able to composite with other polymers to realize multiple functions. Specifically, a gel electrolyte with antifreeze function was accomplished by combining PAM with various polymers (Figure 6.4c). A stable power output under diverse conditions such as hammering and bending was realized. The distinctive point is that it can withstand ultra-low temperatures, even at −20°C, with a capacity retention rate of >90%. Even if it is bent vigorously, it can still maintain structural integrity (Figure 6.4d) [31].

In addition, there are other gel electrolytes based on various polymer matrixes. Pluronic hydrogel electrolyte, which could condense at room temperature and then become liquid again after cooling, can refresh the interface between the electrode and the electrolyte through a simple cooling process when subjected to a strong external mechanical force to restore energy storage and then achieve the purpose of self-healing [32]. Another typical one is a 2D sandwich-like battery prepared with a gel electrolyte composed of poly(N-isopropylacrylamide-co-acrylic acid) with a thermal response function, which can gel the electrolyte at high temperature to inhibit the operation of the battery. When the temperature returns normal, the electrolyte is restored so that the battery performance was revived [33].

6.3 FIBER AQUEOUS ZINC-ION BATTERIES

To mitigate the limitations of the 2D planar structure, 1D fiber aqueous zinc-ion batteries are recognized as a viable approach for further functionality [34,35]. First, the scale of fiber aqueous zinc-ion batteries is reduced to tens to hundreds of microns. It indicates that these fiber aqueous zinc-ion batteries can bear varied deformations, including bending, twisting, winding, and stretching for stable and close body contact. Second, such fiber aqueous zinc-ion batteries can be woven into functional and breathable textiles. On the other hand, matching electrochemical performance with portable and wearable electronic devices is observed. This section reviews the recent developments of fiber aqueous zinc-ion batteries from material synthesis, structural design, and performance majorization. As the crucial components to aqueous zinc-ion batteries, functionalized fiber electrodes are first summarized with the classification of the substrates. A parallel or coaxial structure is applied to construct the full battery on obtaining flexible fiber electrodes, which delivers distinct properties. Furthermore, the working mechanism varies with the electrode and structural design, which is discussed afterward.

6.3.1 Electrodes for Fiber Aqueous Zinc-Ion Batteries

To obtain fiber aqueous zinc-ion batteries, it is crucial to fabricate fiber electrodes that intrinsically bear the deformations mentioned above and ensure high electrical conductivity. Zn metal wires are widely applied directly as the fiber anode for fiber anodes due to their high conductivity and bearable flexibility [36–41]. However, Zn metal is still rigid and heavy, which significantly restrained the flexibility and weavability of the full device.

Polymer fibers are generally lightweight and flexible and are chosen as the substates. PANI and Zn were *in situ* polymerized or electrodeposited on cellulose fiber as the electrodes for fiber aqueous zinc-ion batteries, exhibiting relatively stable power output at various bending angles and after 1,000 bending cycles. However, a conductive coating was introduced due to the limited conductivity of cellulose fibers, indicating the insufficiency of polymer substrate in fiber electrodes [42].

By contrast, carbonaceous materials, such as carbon fibers, CNT fibers, graphene fibers, etc., exhibit lightweight, high surface area, improved tensile strength, enhanced capacitance, and optimal electrical conductivity [43]. Among them, carbon fibers, felts, or cloth were primarily chosen as the substrates for fiber electrodes due to their endurable electrical conductivity and high mechanical strength [37,40,41,44]. However, the electrochemical performance of such fiber electrodes is hindered by the restrained conductivity of carbon fibers and relative rigidity. CNT fibers may be a better choice for developing flexible electrodes due to their superior electrical conductivity, flexibility, and mechanical strength [34,39,41,-44–52]. For example, a well-developed Zn-MnO_2 was transformed into a fibrous shape by electrodepositing MnO_2 on CNT fibers with a diameter of 80–100 μm as the fiber cathode. The MnO_2@CNT fibers were paralleled with Zn wires and constituted fiber aqueous zinc-ion batteries (Figure 6.5a). Owing to the high conductivity and flexibility of CNT

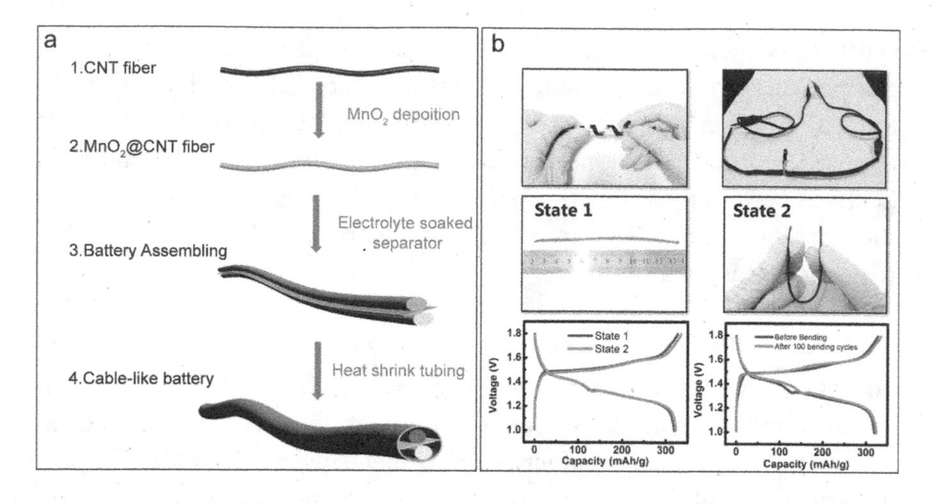

FIGURE 6.5 (a) Schematic illustration of the Zn–MnO_2 cable battery based on MnO_2@CNT (carbon nanotube) fiber cathode and Zn wire anode and (b) its flexibility performance from Ref. [36] with permission of the American Chemical Society.

fibers, the as-prepared fibrous aqueous zinc-ion batteries exhibited a high energy density of 360 Wh·kg^{-1}. They could accommodate deformations such as twisting and bending (Figure 6.5b) [36].

To further enhance the performance of carbonaceous substrates, CNT fibers are further composited with other functional carbon materials. 3D N-doped carbon (NC) nanowall arrays were first introduced to CNT fibers *via* the facile growth of Co-based metal organic framework (MOF) arrays on their surface, followed by calcination and etching. Subsequently, V_2O_5 nanosheets were hydrothermally grown on the carbon skeleton to form the V_2O_5@NC@CNT fiber cathode. The V_2O_5@NC@CNT fiber cathodes were further twisted into fiber aqueous zinc-ion batteries (Figure 6.6a). Competitive rate performance of 47.5% capacity retention at 100-fold current density is observed, thanks to the enhanced ion diffusion and electron transportation of the 3D NC/CNT composites (Figure 6.6b). Besides, the as-prepared fibrous aqueous zinc-ion batteries exhibited steady performance under bending (Figure 6.6b and c) [46]. Other strategies, such as functionalization of CNT fibers by electro-oxidation [47] and compositing CNT fibers with SWCNTs [51], were also developed to enhance the mass diffusion and ion transport for high rate capability.

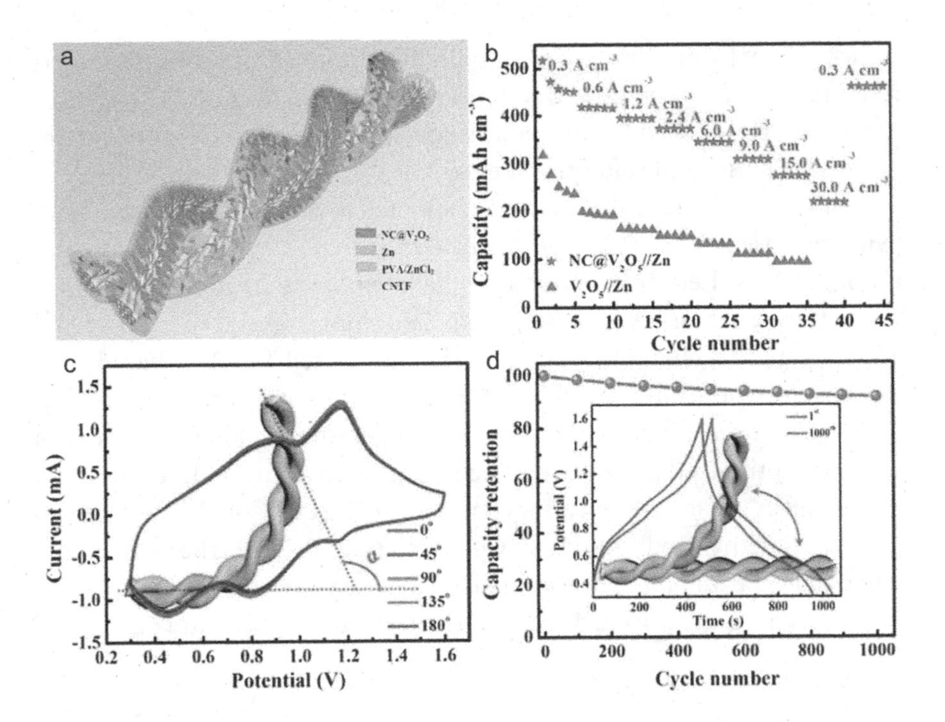

FIGURE 6.6 (a) Schematic illustration of the all-solid-state fiber Zn-ion battery. (b) Comparison of rate capabilities. (c) CV curves at a scan rate of 5 mV·s^{-1} under different bend angles. (d) Galvanostatic charge-discharge curves of the battery during the 1st and 1,000th cycle of bending with an angle of 90°. (Reproduced from Ref. [46] with permission of the Royal Society of Chemistry.)

6.3.2 Electrolyte for Fiber Aqueous Zinc-Ion Batteries

Gel electrolytes are a better substitute for fabricating fiber aqueous zinc-ion batteries due to their high ionic conductivity and self-standing property. As noted before, gel electrolytes can ameliorate the leakage problem, stabilize the electrolyte–electrode interface *via* high viscosity, suppress material dissolution and dendrite growth with restricted H_2O content, and serve as the separator directly [34,53,54].

Generally, gel electrolytes applied in fiber aqueous zinc-ion batteries can be classified as PAM-based [38,45], PVA-based [36,37,39–41,46,49,50,55], gelatin-based [51], and carboxymethyl cellulose-based gel electrolytes [42,47,48]. Designation of the specific combination of Zn salts and polymer matrix is recognized as a viable strategy to endow fiber with unique functions and characteristics. To suppress the Mn dissolution, 2 M $ZnSO_4$+0.1 M $MnSO_4$ aqueous solution was combined with cross-linked

PAM gel to form a waterproof, tailorable, and stretchable electrolyte (Figure 6.7a). The as-synthesized fibrous battery can bear 300% stretching (Figure 6.7b and c) and work stably underwater for 12 h due to the intrinsic stretchability and waterproofness of the PAM gel (Figure 6.7d) [45].

Other polymer networks were also adopted to construct gel electrolytes with purposes. Polyacrylic acid (PAA) was selected as the polymer matrix to build the gel electrolyte due to its larger porous structure and faster mass transfer than PAA gel (Figure 6.8a). Thus, higher capacitance and higher capacity retention at large rates were noted for PAA-based fiber aqueous zinc-ion batteries (Figure 6.8b–d) [44]. To acquire fiber aqueous zinc-ion batteries with a wide operating temperature window, a unique ion gel was synthesized by polymerizing 1-vinyl-3ethylimidazolium dicyanamide and N, N'-methylenebisacrylamide in $Zn(CH_3COO)_2$ dissolved in 1-ethyl-3-methylimidazolium dicyanamide (Figure 6.8e). The as-prepared fiber aqueous zinc-ion batteries delivered steady power output from 0°C to 60°C, which greatly enlarged the working scenarios (Figure 6.8f) [56].

6.3.3 Device Configuration

Typically, the device configuration of fiber batteries can be classified into three species, namely parallel, twisted, and coaxial fiber electrodes [57]. The most straightforward way of transforming fiber aqueous zinc-ion batteries with a 2D sandwich structure into the 1D fibrous structure is the parallel configuration [36,39,44,45,52,55,56]. The parallel configuration resembles the 2D planar structure and shares properties such as short inter-electrode distance and large effective electrode surface area. Therefore, low internal resistance and fast mass transfer are generally realized, facilitating high-performance fiber aqueous zinc-ion batteries. Specifically, a V_6O_{13}@CNT fiber cathode is parallel with another Zn@CNT fiber anode to form a parallel configuration (Figure 6.9a). The air-charging function is realized by the spontaneous oxidization of the discharge product $Zn_{3.49}V_6O_{13}$ to $Zn_{3.49-x}V_6O_{13}$ by O_2 (Figure 6.9b), which is significantly ensured by the large surface area and fast O_2 diffusion endowed by parallel structure (Figure 6.9c) [52]. Furthermore, parallel structure favors integration and scale-up due to the easiness of end-to-end connection.

Twisted configuration is another favorable device layout in fiber aqueous zinc-ion batteries by simply twisting the fiber electrodes together with proper isolation [42,46,47,49–51]. Owing to the similarity between the production of twisted fiber electrodes and textiles, commercial

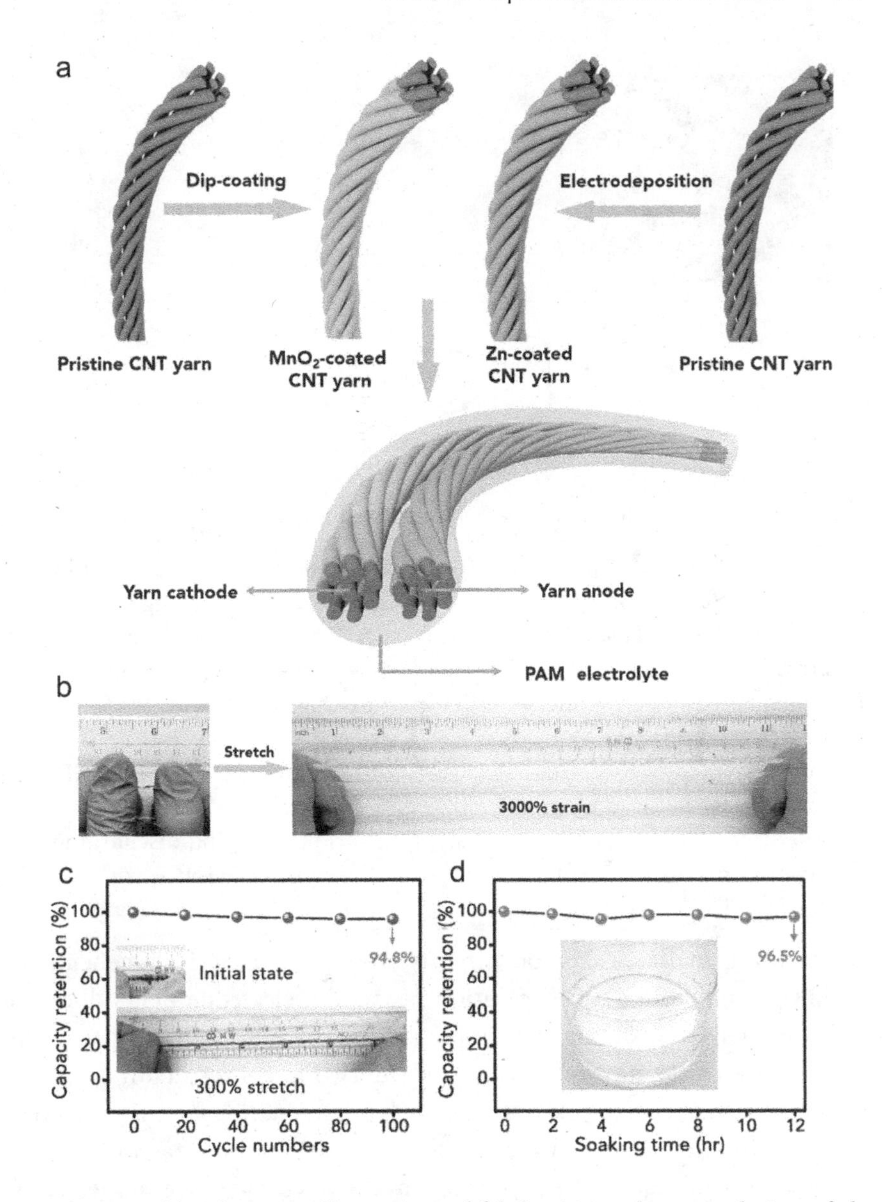

FIGURE 6.7 (a) Schematic illustration of fabrication and encapsulation of the yarn aqueous zinc-ion battery. (b) Relaxed (left) and elongated (right) state of the cross-linked polyacrylamide (PAM), showing good stretchability (3,000% strain). (c) Dependence of capacity retention on cycle numbers with a strain of 300%. (d) Capacity retention test of the yarn aqueous zinc-ion battery for 12 continuous hours of underwater immersion in deionized (DI) water at 24°C, showing superior waterproof ability. (Reproduced from Ref. [45,46] with permission of the American Chemical Society.)

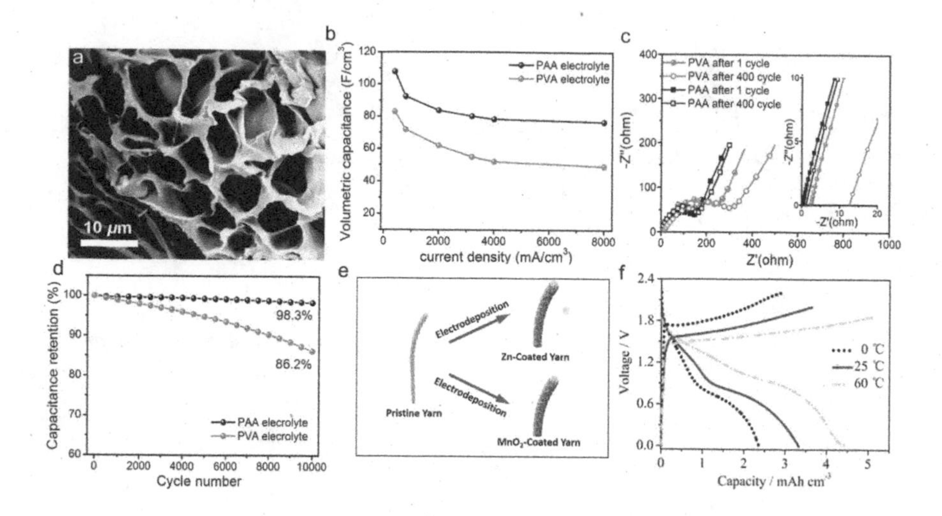

FIGURE 6.8 (a) SEM image of the freeze-dried polyacrylic acid (PAA) hydrogel. (b) The comparison of the rate capability of the ZnFCs using $ZnSO_4$/PAA and $ZnSO_4$/polyvinyl alcohol (PVA) electrolytes. (c) Electrochemical impedance spectroscopy (EIS) Nyquist plots of the ZnFCs using $ZnSO_4$/PAA and $ZnSO_4$/PVA electrolytes after 1 and 400 charging/discharging cycles. (d) The comparison of cycling stability of the ZnFCs using $ZnSO_4$/PAA and $ZnSO_4$/PVA electrolytes. (Reproduced from Ref. [44] with permission of Wiley-VCH.) (e) Schematic illustration of preparation process of fiber electrode. (f) Galvanostatic charge-discharge curves of the yarn battery at various charging/discharging current densities from 13 to 52 mA·cm^{-3}. (Reproduced from Ref. [56] with permission of Elsevier.)

co-spinning facilities can be directly applied, which is beneficial for scale-up production. The fabrication of twisted electrode fibers does not require substrates, stipulating simple processing with light and excellent weavability [57]. However, twisting of the two fibers results in a vulnerable interface and potential damage under deformations. In such a situation, the interface between the twisted electrodes may detach from each other after bending, twisting, winding, and stretching with enlarged resistance, inducing capacity deterioration and battery failure. Furthermore, the unevenly covered gel electrolytes and harsh deformations also lead to short circuit of the fiber batteries [35]. Besides, the inter-electrode distance varies from site to site, impacting the mass transfer kinetics, leading to varied internal resistance under deformation. To obtain uniform electrode distance and better structural integrity, coaxial configuration with multilayer structure is applied in fiber aqueous zinc-ion batteries by depositing

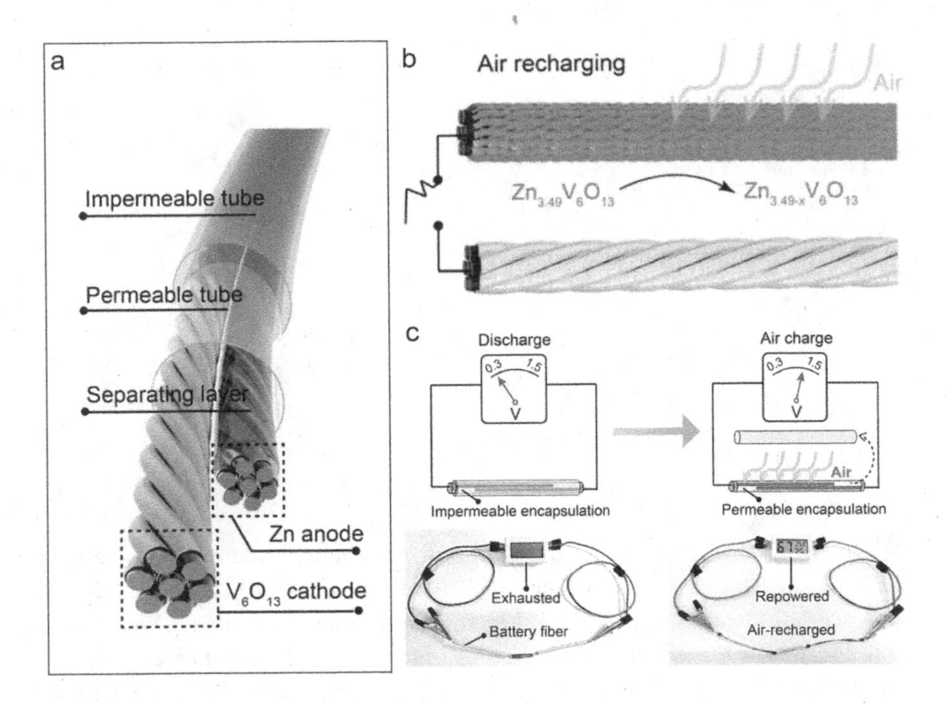

FIGURE 6.9 (a) An illustration of the flexible, double-layer-encapsulated V_6O_{13}/carbon nanotube hybrid fiber (VCF)/Zn battery fiber. (b) Schematic illustration to the VCF/Zn battery fiber during an air-recharging process. (c) A thermometer powered by two air-rechargeable VCF/Zn battery fibers connected in series at exhausted and air-recharged states, respectively. (Reproduced from Ref. [52] with permission of the Royal Society of Chemistry.)

electrode materials and gel electrolytes on core fiber layer by layer. The configuration is generally realized by two modes: (1) Fiber cathode (*i.e.*, MnO_2@CNT) is the core fiber, followed by coating gel electrolyte and Zn foil wrapping [41]. Nevertheless, such a design suffers from probable battery failure due to the robustness loss of outer Zn anode upon cycling due to dissolution. (2) More favorably, Zn wire or Zn@CNT fiber electrode is used directly as the core fiber and anode simultaneously, followed by coating Zn wire with gel electrolytes and outer cathodes in order [37,40]. Due to the well-ordered core-sheath structure of coaxial configuration, functional layers can be introduced to construct a full battery. As shown in Figure 6.10a, a poly(ethylene oxide)$_{53}$-poly(propylene oxide)$_{34}$-poly(-ethylene oxide)$_{53}$ (F77) gel with a unique sol-gel transition feature was introduced between cathode and PAM gel. The fiber aqueous zinc-ion batteries were processed at 0°C to lead the F77 gel to liquidize and re-gelation

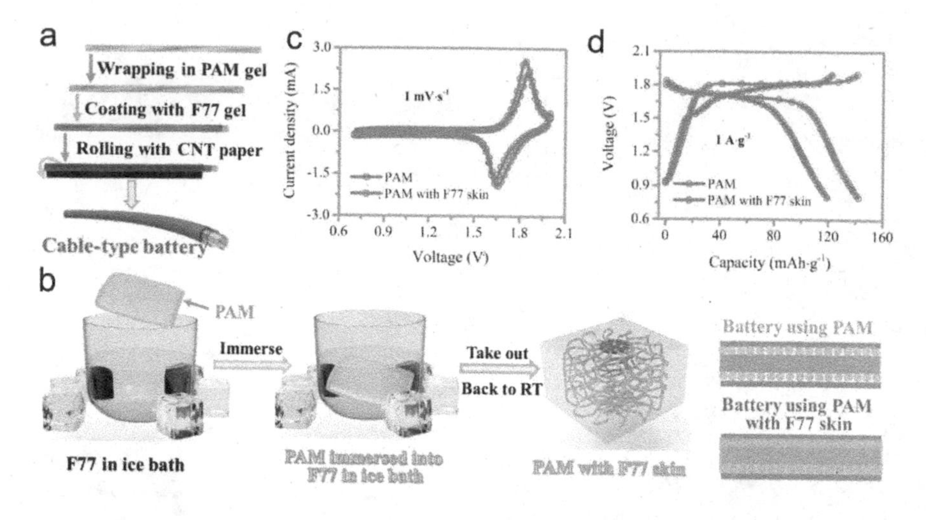

FIGURE 6.10 (a) A schematic illustration for the fabrication of the cable-type $Zn/CoFe(CN)_6$ battery. (b) Schematic illustration of the synthetic process for polyacrylamide (PAM) with F77 skin hydrogel and battery using PAM without and with F77 skin electrolyte. (c) CV curves and (d) galvanostatic charge/discharge curves of cable-type batteries with and without F77 gel, respectively. (Reproduced from Ref. [38] with permission of the Wiley-VCH.)

and form a compact interface with the cathode (Figure 6.10b), obtaining full utilization of cathode materials (Figure 6.10c and d) [38].

All in all, fiber aqueous zinc-ion batteries are an encouraging development trend in fiber energy storage devices due to the intrinsic advantages of aqueous zinc-ion batteries such as low price, high performance, safety and environmental benignity. In the future, proper active material design, gel electrolyte optimization, and structural reconsideration are likely to endow fiber aqueous zinc-ion batteries with higher electrochemical performance, higher flexibility, and more functions, thus bringing fiber aqueous zinc-ion batteries closer to real-life application.

6.4 PERSPECTIVE

Various energy storage devices have been re-designed into flexible and even stretchable configurations in response to the call for portable and wearable electronic devices. With the pursuit of low cost, high electrochemical performance, ensured safety, resource abundance, and environmental friendliness, aqueous zinc-ion batteries, based on the "rock-chair" migration of bivalent Zn ion, are acquiring more attention from academia and industry. Aqueous zinc-ion batteries are transformed into 2D thin-film and 1D fiber structures to satisfy the need for wearable devices. Although advancements have been made, there are still fundamental issues that need to be addressed to better apply flexible aqueous zinc-ion batteries, discussed below.

1. Upon appropriate structural design, flexible anodes (i.e., Zn@CC and Zn@CNT) are developed to replace Zn foil and Zn wires with higher flexibility. Although Zn metal or its carbonaceous composite are regarded as "satisfactory" in neutral or mild acid electrolytes, the anodic mass is usually designed excessively. It results from unavoidable side reactions such as corrosion, passivation, and dendrite growth. Due to the high N/P ratio, the energy and power densities of the full device and the lightweight are sacrificed. Moreover, the anodic parasitic reactions are likely to lead to the short circuit and shortened lifespan of flexible aqueous zinc-ion batteries. Strategies such as surface modifications, structural design, and electrolyte optimizations have been applied to Zn anode [58]. It is thus crucial to adapt these strategies to fiber aqueous zinc-ion batteries.
2. Despite Zn metal's high volumetric and mass-specific density, the paired cathode material generally exhibits much lower specific capacity, which has become the shortboard in developing real-life applications. Regarding this issue, two pathways may be considered. First, the areal mass loading of the cathode material is ordinarily low in lab research (<5 $mg{\cdot}cm^{-2}$) compared with commercial requirements (>10 $mg{\cdot}cm^{-2}$). On the one hand, high loading mass leads to low utilization of cathode material and subsequent low electrochemical performance. On the other hand, stacked cathode material significantly reduces the flexibility of the device. Therefore, one viable strategy is designing an appropriate substrate endowed with high surface area and flexibility. Second, the pursuit of cathode materials with high performance is eternal. Defect engineering, heterostructure, cation doping, and compositing have been proven effective. Novel techniques must be developed for cathode materials with high energy/power density and long cycle life.

REFERENCES

1. Cai, Z. Ou, Y. Wang, J. Xiao, R. Fu, L. Yuan, Z. Zhan, R. Sun, Y. 2020. Chemically resistant Cu–Zn/Zn composite anode for long cycling aqueous batteries. *Energy Storage Materials* 27: 205–211.
2. Liang, P. Yi, J. Liu, X. Wu, K. Wang, Z. Cui, J. Liu, Y. Wang, Y. Xia, Y. Zhang, J. 2020. Highly reversible Zn anode enabled by the controllable formation of nucleation sites for Zn-based batteries. *Advanced Functional Materials* 30: 1908528.
3. Li, M. He, Q. Li, Z. Li, Q. Zhang, Y. Meng, J. Liu, X. Li, S. Wu, B. Chen, L. 2019. A novel dendrite-free Mn^{2+}/Zn^{2+} hybrid battery with 2.3 V voltage window and 11000-cycle lifespan. *Advanced Energy Materials* 9: 1901469.
4. Hao, J. Li, X. Zhang, S. Yang, F. Zeng, X. Zhang, S. Bo, G. Wang, C. Guo, Z. 2020. Designing dendrite-free zinc anodes for advanced aqueous zinc batteries. *Advanced Functional Materials* 30: 2001263.
5. Zeng, Y. Zhang, X. Qin, R. Liu, X. Fang, P. Zheng, D. Tong, Y. Lu, X. 2019. Dendrite-free zinc deposition induced by multifunctional CNT frameworks for stable flexible Zn-ion batteries. *Advanced Materials* 31: 1903675.
6. Shi, X. Xu, G. Liang, S. Li, C. Guo, S. Xie, X. Ma, X. Zhou, J. 2019. Homogeneous deposition of zinc on three-dimensional porous copper foam as a superior zinc metal anode. *ACS Sustainable Chemistry & Engineering* 7: 17737–17746.
7. Jia, X. Liu, C. Neale, Z. G. Yang, J. Cao, G. 2020. Active materials for aqueous zinc ion batteries: synthesis, crystal structure, morphology, and electrochemistry. *Chemical Reviews* 120: 7795–7866.
8. Huang, J. Wang, Z. Hou, M. Dong, X. Liu, Y. Wang, Y. Xia, Y. 2018. Polyaniline-intercalated manganese dioxide nanolayers as a high-performance cathode material for an aqueous zinc-ion battery. *Nature Communications* 9: 1–8.
9. Alfaruqi, M. H. Islam, S. Mathew, V. Song, J. Kim, S. Tung, D. P. Jo, J. Kim, S. Baboo, J. P. Xiu, Z. 2017. Ambient redox synthesis of vanadium-doped manganese dioxide nanoparticles and their enhanced zinc storage properties. *Applied Surface Science* 404: 435–442.
10. Zhang, Y. Deng, S. Luo, M. Pan, G. Zeng, Y. Lu, X. Ai, C. Liu, Q. Xiong, Q. Wang, X. 2019. Defect promoted capacity and durability of N-MnO_{2-x} branch arrays via low-temperature NH_3 treatment for advanced aqueous zinc ion batteries. *Small* 15: 1905452.
11. Xu, D. Li, B. Wei, C. He, Y.-B. Du, H. Chu, X. Qin, X. Yang, Q.-H. Kang, F. 2014. Preparation and characterization of MnO_2/acid-treated CNT nanocomposites for energy storage with zinc ions. *Electrochimica Acta* 133: 254–261.
12. Wang, J. Wang, J.-G. Liu, H. You, Z. Wei, C. Kang, F. 2019. Electrochemical activation of commercial MnO microsized particles for high-performance aqueous zinc-ion batteries. *Journal of Power Sources* 438: 226951.

13. Zhang, N. Dong, Y. Jia, M. Bian, X. Wang, Y. Qiu, M. Xu, J. Liu, Y. Jiao, L. Cheng, F. 2018. Rechargeable aqueous Zn–V_2O_5 battery with high energy density and long cycle life. *ACS Energy Letters* 3: 1366–1372.
14. Yan, M. He, P. Chen, Y. Wang, S. Wei, Q. Zhao, K. Xu, X. An, Q. Shuang, Y. Shao, Y. 2018. Water-lubricated intercalation in $V_2O_5 \cdot nH_2O$ for high-capacity and high-rate aqueous rechargeable zinc batteries. *Advanced Materials* 30: 1703725.
15. He, P. Quan, Y. Xu, X. Yan, M. Yang, W. An, Q. He, L. Mai, L. 2017. High-performance aqueous zinc–ion battery based on layered $H_2V_3O_8$ nanowire cathode. *Small* 13: 1702551.
16. Hu, P. Zhu, T. Wang, X. Wei, X. Yan, M. Li, J. Luo, W. Yang, W. Zhang, W. Zhou, L. 2018. Highly durable $Na_2V_6O_{16} \cdot 1.63H_2O$ nanowire cathode for aqueous zinc-ion battery. *Nano Letters* 18: 1758–1763.
17. Paolella, A. Faure, C. Timoshevskii, V. Marras, S. Bertoni, G. Guerfi, A. Vijh, A. Armand, M. Zaghib, K. 2017. A review on hexacyanoferrate-based materials for energy storage and smart windows: challenges and perspectives. *Journal of Materials Chemistry A* 5: 18919–18932.
18. Zhang, L. Chen, L. Zhou, X. Liu, Z. 2015. Towards high-voltage aqueous metal-ion batteries beyond 1.5 V: the zinc/zinc hexacyanoferrate system. *Advanced Energy Materials* 5: 1400930.
19. Zhang, N. Cheng, F. Liu, Y. Zhao, Q. Lei, K. Chen, C. Liu, X. Chen, J. 2016. Cation-deficient spinel $ZnMn_2O_4$ cathode in Zn $(CF_3SO_3)_2$ electrolyte for rechargeable aqueous Zn-ion battery. *Journal of the American Chemical Society* 138: 12894–12901.
20. Huang, S. Wan, F. Bi, S. Zhu, J. Niu, Z. Chen, J. 2019. A self-healing integrated all-in-one zinc-ion battery. *Angewandte Chemie* 131: 4357–4361.
21. Wang, T. Li, C. Xie, X. Lu, B. He, Z. Liang, S. Zhou, J. 2020. Anode materials for aqueous zinc ion batteries: mechanisms, properties, and perspectives. *ACS Nano* 14: 16321–16347.
22. Guo, Z. Ma, Y. Dong, X. Huang, J. Wang, Y. Xia, Y. 2018. An environmentally friendly and flexible aqueous zinc battery using an organic cathode. *Angewandte Chemie International Edition* 57: 11737–11741.
23. Wan, F. Zhang, L. Dai, X. Wang, X. Niu, Z. Chen, J. 2018. Aqueous rechargeable zinc/sodium vanadate batteries with enhanced performance from simultaneous insertion of dual carriers. *Nature Communications* 9: 1–11.
24. Wang, Z. Ruan, Z. Ng, W. S. Li, H. Tang, Z. Liu, Z. Wang, Y. Hu, H. Zhi, C. 2018. Integrating a triboelectric nanogenerator and a zinc-ion battery on a designed flexible 3D spacer fabric. *Small Methods* 2: 1800150.
25. Chao, D. Zhu, C. Song, M. Liang, P. Zhang, X. Tiep, N. H. Zhao, H. Wang, J. Wang, R. Zhang, H. 2018. A high-rate and stable quasi-solid-state zinc-ion battery with novel 2D layered zinc orthovanadate array. *Advanced Materials* 30: 1803181.
26. Wan, F. Huang, S. Cao, H. Niu, Z. 2020. Freestanding potassium vanadate/carbon nanotube films for ultralong-life aqueous zinc-ion batteries. *ACS Nano* 14: 6752–6760.

27. Shi, J. Wang, S. Chen, X. Chen, Z. Du, X. Ni, T. Wang, Q. Ruan, L. Zeng, W. Huang, Z. 2019. An ultrahigh energy density quasi-solid-state zinc ion microbattery with excellent flexibility and thermostability. *Advanced Energy Materials* 9: 1901957.
28. Ma, L. Chen, S. Li, N. Liu, Z. Tang, Z. Zapien, J. A. Chen, S. Fan, J. Zhi, C. 2020. Hydrogen-free and dendrite-free all-solid-state Zn-ion batteries. *Advanced Materials* 32: 1908121.
29. Ma, L. Chen, S. Li, X. Chen, A. Dong, B. Zhi, C. 2020. Liquid-free all-solid-state zinc batteries and encapsulation-free flexible batteries enabled by in situ constructed polymer electrolyte. *Angewandte Chemie International Edition* 132: 24044–24052.
30. Li, H. Han, C. Huang, Y. Huang, Y. Zhu, M. Pei, Z. Xue, Q. Wang, Z. Liu, Z. Tang, Z. 2018. An extremely safe and wearable solid-state zinc ion battery based on a hierarchical structured polymer electrolyte. *Energy & Environmental Science* 11: 941–951.
31. Mo, F. Liang, G. Meng, Q. Liu, Z. Li, H. Fan, J. Zhi, C. 2019. A flexible rechargeable aqueous zinc manganese-dioxide battery working at– 20°C. *Energy & Environmental Science* 12: 706–715.
32. Zhao, J. Sonigara, K. K. Li, J. Zhang, J. Chen, B. Zhang, J. Soni, S. S. Zhou, X. Cui, G. Chen, L. 2017. A smart flexible zinc battery with cooling recovery ability. *Angewandte Chemie International Edition* 56: 7871–7875.
33. Mo, F. Li, H. Pei, Z. Liang, G. Ma, L. Yang, Q. Wang, D. Huang, Y. Zhi, C. 2018. A smart safe rechargeable zinc ion battery based on sol-gel transition electrolytes. *Science Bulletin* 63: 1077–1086.
34. Yu, P. Zeng, Y. Zhang, H. Yu, M. Tong, Y. Lu, X. 2019. Flexible Zn-ion batteries: recent progresses and challenges. *Small* 15: 1804760.
35. Sun, H. Zhang, Y. Zhang, J. Sun, X. Peng, H. 2017. Energy harvesting and storage in 1D devices. *Nature Reviews Materials* 2: 1–12.
36. Wang, K. Zhang, X. Han, J. Zhang, X. Sun, X. Li, C. Liu, W. Li, Q. Ma, Y. 2018. High-performance cable-type flexible rechargeable Zn battery based on MnO_2@CNT fiber microelectrode. *ACS Applied Materials & Interfaces* 10: 24573–24582.
37. Wan, F. Zhang, L. Wang, X. Bi, S. Niu, Z. Chen, J. 2018. An aqueous rechargeable zinc-organic battery with hybrid mechanism. *Advanced Functional Materials* 28: 1804975.
38. Ma, L. Chen, S. Long, C. Li, X. Zhao, Y. Liu, Z. Huang, Z. Dong, B. Zapien, J. A. Zhi, C. 2019. Achieving high-voltage and high-capacity aqueous rechargeable zinc ion battery by incorporating two-species redox reaction. *Advanced Energy Materials* 9: 1902446.
39. Wang, C. He, T. Cheng, J. Guan, Q. Wang, B. 2020. Bioinspired interface design of sewable, weavable, and washable fiber zinc batteries for wearable power textiles. *Advanced Functional Materials* 30: 2004430.

40. Zhao, Y. Wang, Y. Zhao, Z. Zhao, J. Xin, T. Wang, N. Liu, J. 2020. Achieving high capacity and long life of aqueous rechargeable zinc battery by using nanoporous-carbon-supported poly (1, 5-naphthalenediamine) nanorods as cathode. *Energy Storage Materials* 28: 64–72.
41. Fan, W. Ding, J. Ding, J. Zheng, Y. Song, W. Lin, J. Xiao, C. Zhong, C. Wang, H. Hu, W. 2021. Identifying heteroatomic and defective sites in carbon with dual-ion adsorption capability for high energy and power zinc ion capacitor. *Nano-Micro Letters* 13: 1–18.
42. Yi, H. Ma, Y. Zhang, S. Na, B. Zeng, R. Zhang, Y. Lin, C. 2019. Robust Aqueous Zn-ion fiber battery based on high-strength cellulose yarns. *ACS Sustainable Chemistry & Engineering* 7: 18894–18900.
43. Meng, Y. Zhao, Y. Hu, C. Cheng, H. Hu, Y. Zhang, Z. Shi, G. Qu, L. 2013. All-graphene core-sheath microfibers for all-solid-state, stretchable fibriform supercapacitors and wearable electronic textiles. *Advanced Materials* 25: 2326–2331.
44. Zhang, X. Pei, Z. Wang, C. Yuan, Z. Wei, L. Pan, Y. Mahmood, A. Shao, Q. Chen, Y. 2019. Flexible zinc-ion hybrid fiber capacitors with ultrahigh energy density and long cycling life for wearable electronics. *Small* 15: 1903817.
45. Li, H. Liu, Z. Liang, G. Huang, Y. Huang, Y. Zhu, M. Pei, Z. Xue, Q. Tang, Z. Wang, Y. 2018. Waterproof and tailorable elastic rechargeable yarn zinc ion batteries by a cross-linked polyacrylamide electrolyte. *ACS Nano* 12: 3140–3148.
46. He, B. Zhou, Z. Man, P. Zhang, Q. Li, C. Xie, L. Wang, X. Li, Q. Yao, Y. 2019. V_2O_5 nanosheets supported on 3D N-doped carbon nanowall arrays as an advanced cathode for high energy and high power fiber-shaped zinc-ion batteries. *Journal of Materials Chemistry A* 7: 12979–12986.
47. Pan, Z. Yang, J. Yang, J. Zhang, Q. Zhang, H. Li, X. Kou, Z. Zhang, Y. Chen, H. Yan, C. 2019. Stitching of $Zn_3(OH)_2V_2O_7{\cdot}2H_2O$ 2D nanosheets by 1D carbon nanotubes boosts ultrahigh rate for wearable quasi-solid-state zinc-ion batteries. *ACS Nano* 14: 842–853.
48. Zhang, Q. Li, C. Li, Q. Pan, Z. Sun, J. Zhou, Z. He, B. Man, P. Xie, L. Kang, L. 2019. Flexible and high-voltage coaxial-fiber aqueous rechargeable zinc-ion battery. *Nano Letters* 19: 4035–4042.
49. Li, Q. Zhang, Q. Liu, C. Zhou, Z. Li, C. He, B. Man, P. Wang, X. Yao, Y. 2019. Anchoring V_2O_5 nanosheets on hierarchical titanium nitride nanowire arrays to form core–shell heterostructures as a superior cathode for high-performance wearable aqueous rechargeable zinc-ion batteries. *Journal of Materials Chemistry A* 7: 12997–13006.
50. Liu, C. Li, Q. Sun, H. Wang, Z. Gong, W. Cong, S. Yao, Y. Zhao, Z. 2020. MOF-derived vertically stacked Mn_2O_3@C flakes for fiber-shaped zinc-ion batteries. *Journal of Materials Chemistry A* 8: 24031–24039.

51. Lin, Y. Zhou, F. Xie, M. Zhang, S. Deng, C. 2020. $V_6O_{13-\delta}$@C nanoscrolls with expanded distances between adjacent shells as a high-performance cathode for a knittable zinc-ion battery. *ChemSusChem* 13: 3696–3706.
52. Liao, M. Wang, J. Ye, L. Sun, H. Li, P. Wang, C. Tang, C. Cheng, X. Wang, B. Peng, H. 2021. A high-capacity aqueous zinc-ion battery fiber with air-recharging capability. *Journal of Materials Chemistry A* 9: 6811–6818.
53. Liao, M. Ye, L. Zhang, Y. Chen, T. Peng, H. 2019. The recent advance in fiber-shaped energy storage devices. *Advanced Electronic Materials* 5: 1800456.
54. Liu, Z. Mo, F. Li, H. Zhu, M. Wang, Z. Liang, G. Zhi, C. 2018. Advances in flexible and wearable energy-storage textiles. *Small Methods* 2: 1800124.
55. Liu, Y. Wang, J. Zeng, Y. Liu, J. Liu, X. Lu, X. 2020. Interfacial engineering coupled valence tuning of MoO_3 cathode for high-capacity and high-rate fiber-shaped zinc-ion batteries. *Small* 16: 1907458.
56. Liu, J. Nie, N. Wang, J. Hu, M. Zhang, J. Li, M. Huang, Y. 2020. Initiating a wide-temperature-window yarn zinc ion battery by a highly conductive iongel. *Materials Today Energy* 16: 100372.
57. Zhou, Y. Wang, C. H. Lu, W. Dai, L. 2020. Recent advances in fiber-shaped supercapacitors and lithium-ion batteries. *Advanced Materials* 32: 1902779.
58. Cao, Z. Zhuang, P. Zhang, X. Ye, M. Shen, J. Ajayan, P. M. 2020. Strategies for dendrite-free anode in aqueous rechargeable zinc ion batteries. *Advanced Energy Materials* 10: 2001599.

CHAPTER 7

Flexible Lithium-Air Batteries

7.1 OVERVIEW OF LITHIUM-AIR BATTERIES

7.1.1 General Principle

As schematically shown in Figure 7.1, a typical lithium-air battery consists of a Li metal anode, a porous air cathode with oxygen reduction reaction (ORR)/oxygen evolution reaction (OER) catalytic materials, and an aprotic electrolyte made of an aprotic solvent dissolved with a Li salt [1]. The general reaction steps of lithium-air batteries can be understood through the following equations 7.1–7.6:

Li anode:

$$Li \rightarrow Li^{+} + e^{-} \quad (7.1)$$

Air cathode:

$$O_2 + e^{-} \rightarrow O_2^{-} \quad (7.2)$$

$$O_2^{-} + Li^{+} \rightarrow LiO_2 \quad (7.3)$$

$$LiO_2 + Li^{+} + e^{-} \rightarrow Li_2O_2 \quad (7.4)$$

$$2LiO_2 \rightarrow Li_2O_2 + O_2 \quad (7.5)$$

DOI: 10.1201/9781003273677-7

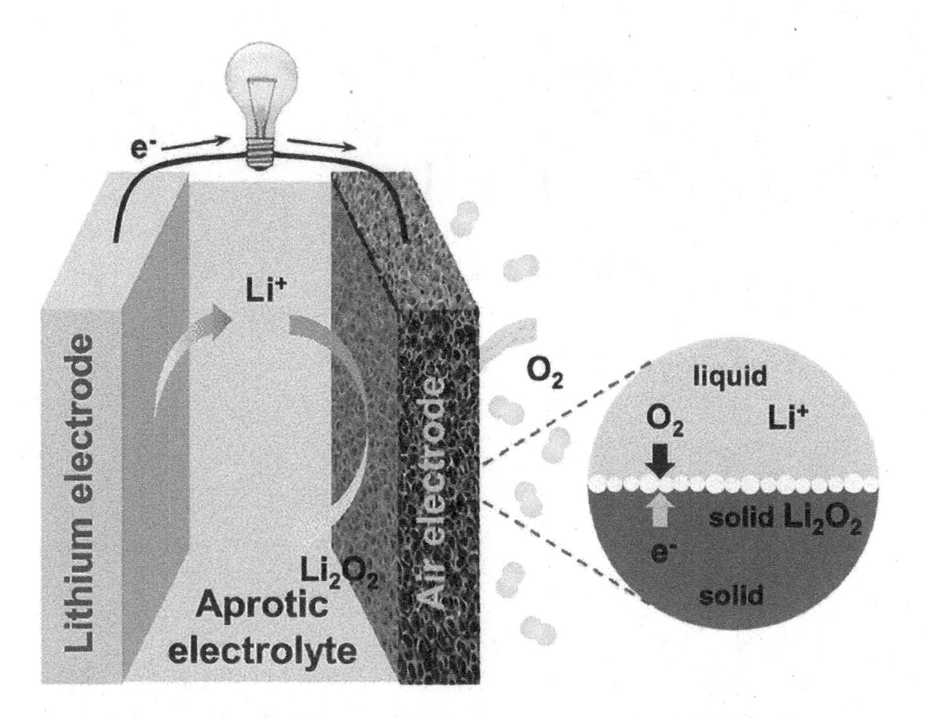

FIGURE 7.1 Schematic operation principle of lithium-air batteries. (Reproduced from Ref. [1] with permission of the Royal Society of Chemistry.)

Overall:

$$2Li + O_2 \rightarrow Li_2O_2 \left(E = 2.96\ V\right) \tag{7.6}$$

During discharging, Li metal is oxidized at the electrode, emitting electrons and Li^+ while atmospheric O_2 dissolves first into the aprotic electrolyte and then reduces through the one-electron-transfer ORR at the two-phase boundaries between the electrolyte (liquid) and the catalytic material (solid) to form O_2^- [2]. The O_2^- combined with Li^+ is transported *via* the electrolyte to form LiO_2 on the cathode. Due to the inherent thermodynamic instability of LiO_2, it spontaneously converts into Li_2O_2 through further disproportionation reaction and deposits on the surface of the air electrode. Upon charging, Li_2O_2 is decomposed into Li^+ and O_2 under the assistance of the OER sites.

Notably, due to the high sensitivity of Li metal to air and moisture, the fabrication processes of lithium-air batteries are more complicated and rigorous. All the assembling processes should be operated in Ar-protected glove boxes with harsh requirements regarding the H_2O and O_2 contents (commonly less than 1 ppm). Furthermore, considering the formation of LiO_2 and Li_2O_2, most lithium-air batteries have to be evaluated in pure O_2 atmospheres [3,4]. Otherwise, CO_2, N_2, H_2O, and other air components can easily react with the discharge products, leading to a faster decay of the cycling performance or even sudden shutting down of the battery. Therefore, lithium-air batteries are supposed to be technically described as lithium-oxygen batteries.

Despite these harsh preparations and working conditions, lithium-air batteries are still sufficiently captivating for both academy and industry because of their unique electrochemistry and convenient O_2-harvesting ability from the ambient environment. More encouragingly, through coupling multifunctional catalyst components and designing a gas protective layer, lithium-oxygen batteries have recently embarked on the way to becoming real lithium-air batteries that may work in ambient air [5,6].

7.1.2 Cathodes

The air cathode is a significant component in lithium-air batteries, which provides the reaction places for both ORR and OER. A typical air cathode is commonly made of a porous current collector and catalyst layer. The current collector provides physical and conductive support for the catalyst and a pathway for oxygen diffusion during discharge and charge processes. The basic requirements include high electrical conductivity, high mechanical stability, and excellent air permeability. The catalyst layer is composed of bifunctional catalysts, carbon conductive matrix, and binders [7,8]. Among them, the bifunctional catalysts possess high catalytic activity to facilitate the ORR and OER reactions of the lithium-air battery.

In general, all components in flexible lithium-air batteries should be deformable to accommodate stress and strain in deformation. Due to high flexibility, large surface area, and facile processing, carbon-based materials (e.g., carbon textile, carbon fiber, and aligned carbon nanotube (CNT) sheet) are preferred for use as flexible current collectors. For example, a flexible and recoverable air cathode constructed on carbon textile by depositing densely TiO_2 nanowire arrays (NAs) is proposed to make a flexible lithium-air battery (Figure 7.2a) [9]. Though fully covered with TiO_2 NAs, the obtained air cathode still inherited the flexible characteristic of the carbon textile. The TiO_2 NA/carbon textile demonstrated high flexibility and excellent recoverability, which can be restored after washing. Significantly, the recoverable feature can extend the cycling life and

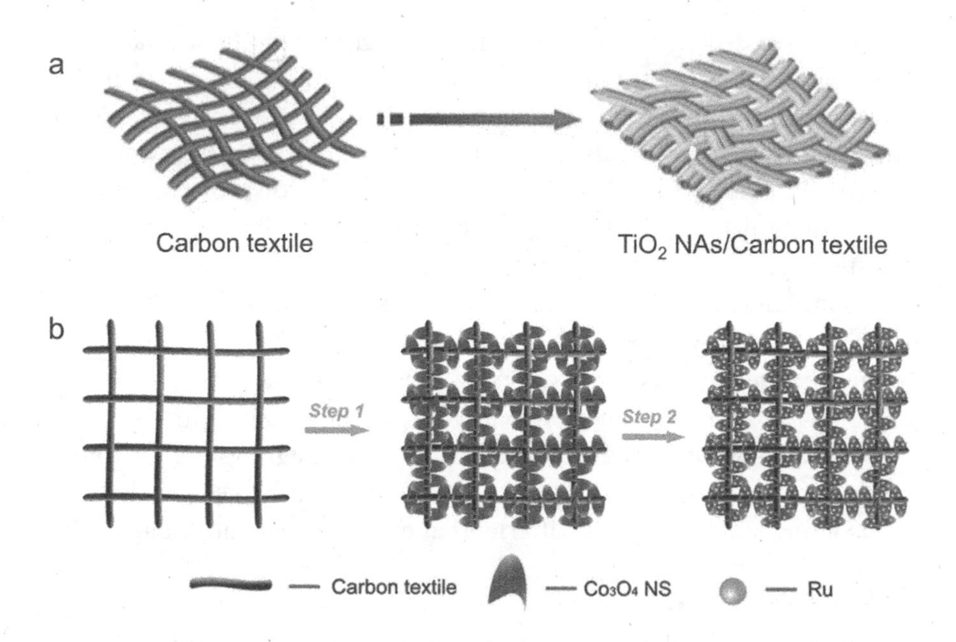

FIGURE 7.2 (a) Schematic representation for the design of a free-standing cathode based on carbon textile. Reproduced from Ref. [9] with permission of Nature Publishing Group. (b) Schematic representation for the design of a binder-free cathode based on carbon textile. (Reproduced from Ref. [10] with permission of Wiley-VCH.)

reduce the cost of the flexible lithium-air battery. Afterward, another flexible air cathode based on the carbon textile and Co_3O_4 nanosheet (NS) arrays is reported (Figure 7.2b) [10]. First, the Co_3O_4 NS arrays were grown on carbon textile by electrodeposition. Second, it was decorated with Ru nanoparticles by the impregnation and reduction method. The 3D hierarchical structure and the uniform distribution of Co_3O_4 NS-Ru on carbon textile resulted in nanosheet-shaped Li_2O_2 growth and improved electrochemical performance of the battery. A flexible lithium-air battery was finally assembled using the flexible Co_3O_4 NS-Ru/carbon textile cathode, demonstrating the application potential of carbon textile.

Although carbon materials have been extensively studied due to their lightweight, low cost, and relatively high catalytic performance, carbon corrosion is a challenging issue during battery operation [11]. In lithium-air batteries, carbon materials can chemically react with oxidative products (e.g., LiO_2 and Li_2O_2) [12]. Especially, carbon decomposes during charging when the charge voltage is above 3.5 V and actively promotes electrolyte decomposition on discharge and charge processes, leading to cathode passivation and premature cell death [13,14]. Metal materials, e.g., nickel foam, stainless-steel mesh, and titanium mesh, have recently been used for lithium-air batteries due to their higher electrical conductivities and electrochemically oxidative stabilities [15–18]. However, the larger densities of metal substrates inevitably lower the whole battery's practical specific capacities and energy densities [19]. Therefore, a compromise must be considered between cathode stability and energy density in developing lithium-air batteries.

In addition to the current collector, the bifunctional catalyst is also essential and can facilitate the ORR and OER reactions of the lithium-air battery during the charging and discharging processes. Although many carbon-based current collectors have shown catalytic ORR activity, carbon's relatively low catalytic activity in the OER remains a problem. Therefore, considerable research efforts have been devoted to searching for powerful electrocatalysts that can accelerate the reaction kinetics associated with the reversible dissociation of discharge product, e.g., Li_2O_2. The chemical tuning of the carbon surface property using heteroatoms (e.g., N, S, O, or P) has effectively improved the ORR/OER activity [20–23]. It has been computationally and experimentally suggested that the electrocatalytic activity of nanostructured carbon materials toward oxygen electrochemistry is improved after heteroatom doping due to the interactions between the lone pair electrons of heteroatoms and π-systems of carbon [22,24]. Nitrogen doping is particularly attractive, thanks to its strong electron affinity and substantially high positive charge density of the adjacent C atoms. For instance, a powerful electrocatalyst of vertically aligned nitrogen-containing CNTs showed much better electrocatalytic activity, long-term operation stability, and tolerance to crossover effect, which is verified experimentally [21]. Besides heteroatom-doped carbon materials, noble metals (e.g., Pt, Pd, Au, Ru, and Ir), noble metals oxides (e.g., RuO_2 and IrO_2), and transitional metal oxides (Co_3O_4, MnO_2, NiO, $MnCo_2O_4$, $NiCo_2O_4$, $CuCo_2O_4$, and $ZnCo_2O_4$) have also been surging due to their excellent electrocatalytic activity for both ORR and OER [25–28]. For example, a bifunctional Pt/Au catalyst is prepared by loading PtAu nanoparticles on carbon, and then the catalyst and separator (Celgard C480) are combined as the air cathode of a lithium-air battery [29]. As a result, both charge and discharge overpotentials of the lithium-air battery were reduced. As another example, a transitional metal oxide such as $ZnCo_2O_4$ has also aroused much interest due to its good electronic conductivity and catalytic activity. A three-dimensionally ordered mesoporous $ZnCo_2O_4$ material was synthesized *via* a hard template and used as a bifunctional electrocatalyst for the lithium-air batteries [30]. The reversibility and cycling stability of the resulting lithium-air batteries was improved remarkably.

It is worth noting that the limited contact area between the solid catalyst as summarized above and the deposited Li_2O_2 limits the electron flow during the OER process, thus generally leading to higher overcharge potential and lower energy efficiency. More recently, the use of a soluble redox mediator (RM) as catalyst was proposed to circumvent this issue [31]. The diffusible RM can quickly move around in the electrolyte and had wet contact with solid discharge products of Li_2O_2, thus maximizing the interaction areas for the catalysis, which led to a marked increase in the energy efficiency and cycle life of lithium-air batteries. Accordingly, researchers have reported several types of soluble catalysts, such as LiI, LiBr, tetrathiafulvalene (TTF), tetramethylpiperidinyloxyl (TEMPO), and iron phthalocyanine (FePc), all of which have exhibited significantly reduced overpotentials of lithium-air batteries [25,32–34]. Figure 7.3a illustrates the reaction mechanism of the RM in lithium-air batteries [31]. First, the RM is oxidized during a charging process near the air cathode surface (Step 1). Second, the oxidized RM (RM^+) freely diffuses in the electrolyte and chemically reacts with Li_2O_2, decomposing it into $2Li^+$ and O_2, and reduces back to the initial state of the RM (Step 2). Because the

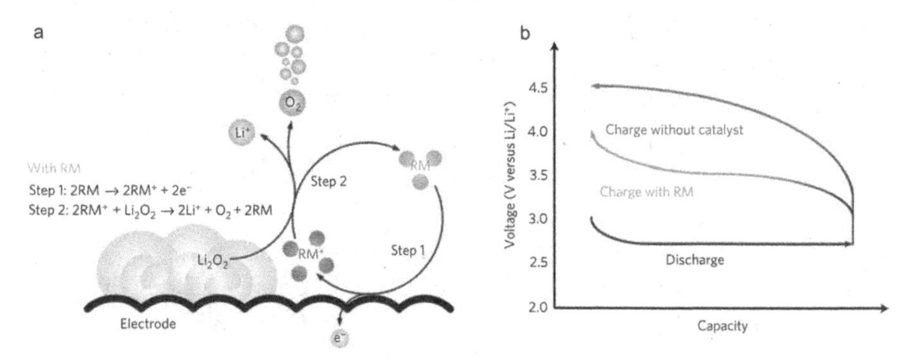

FIGURE 7.3 (a) Schematic illustration of the reaction mechanism of redox mediator (RM) for lithium-air batteries. (b) Schematic discharge and charge profiles of lithium-air batteries with and without RM. (Reproduced from Ref. [31] with permission of Nature Publishing Group.)

charging voltage of the cell is determined by Step 1 (that is, electrochemical oxidation of the RM), charge polarization can be reduced by choosing an RM with a suitable redox potential (Figure 7.3b).

7.1.3 Anodes

In the current research of flexible lithium-air batteries, the anodes are usually Li foils, Li ribbons, and Li sticks/wires, possessing flexible properties to some extent and being used directly in flexible lithium-air batteries. However, because of the semi-open structure of the lithium-air battery, the Li metal anode is likely to react with water, O_2, and electrolyte in the cycling process. After continuous desorption and deposition, the surface of the Li metal anode is prone to powdering and forming an uneven surface, which seriously affects the performance of the anode. Besides, the flexibility of Li metal is limited, and it is not easy to meet higher requirements of deformation, e.g., stretching. Many research efforts were also devoted to improving the performance and safety of metal Li anodes in lithium metal batteries through various strategies, such as applying artificial solid electrolyte interphase (SEI) layers, solid-state membranes, high-concentration electrolytes, or replacing the Li metal with other Li-containing materials [35]. For example, a Li-ZnO@CNT fiber anode was fabricated by infusing a suitable amount of molten Li into ZnO nanowire array-modified CNT fiber (Figure 7.4a) [36]. Because ZnO is lithiophilic, Li prefers to deposit as a film coating rather than combined with the CNT fiber. Hence, the obtained hybrid anode inherited the superior flexibility of pristine CNT fiber (Figure 7.4b). As another example, a stretchable Li array electrode was reported based on a stretchable copper (Cu) current collector, as schematically shown in Figure 7.4c [37]. First, the Cu springs made of Cu wire were connected in series by Cu sheets. Second, Li sheets were paved on the Cu sheets point-to-point and covered by a punched Ecoflex film. This stretchable metal electrode associated with a gel electrolyte and a rippled air electrode enabled good stretchability of the battery, which could maintain stability under strains up to 100% (Figure 7.4d).

7.1.4 Electrolytes

The electrolyte is a medium for Li^+ ion and O_2 transportation during the process of battery operation. An electrolyte used in lithium-air batteries should have the following features: (1) high stability in O_2-rich electrochemical conditions; (2) low viscosity, supporting fast Li^+ ion transport;

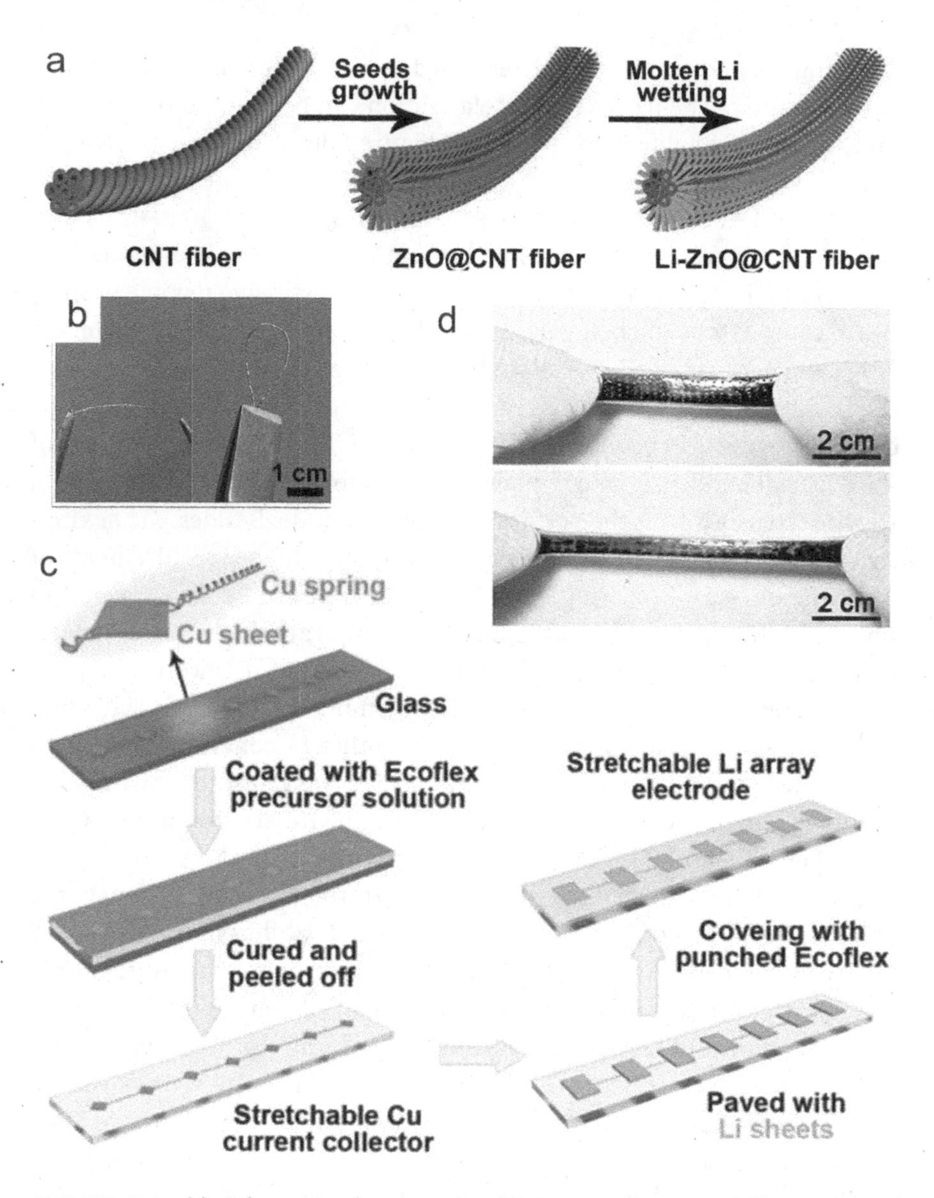

FIGURE 7.4 (a) Schematic showing the fabrication process of fiber Li metal anode. (b) Photographs of the fiber Li metal anode under bending conditions. Reproduced from Ref. [36] with permission of Elsevier. (c) Schematic illustration to the fabrication of stretchable Li array electrode. (d) Photographs of the stretchable lithium-air battery before and after stretching. (Reproduced from Ref. [37] with permission of the Royal Society of Chemistry.)

(3) high solubility of Li salt; and (4) nontoxicity and economic use [38]. In addition, for flexible lithium-air batteries, the electrolyte must be able to endure various deformation modes. Otherwise, it may cause safety issues. With the development of lithium-air batteries, a series of electrolytes has been exploited. Electrolytes can generally be categorized into two types according to their appearance: liquid and solid electrolytes.

According to different operation systems, liquid electrolytes are generally divided into aqueous and nonaqueous electrolytes. Flexible lithium-air batteries are mainly nonaqueous electrolytes to avoid the reactions between Li metal and water. The commonly operated nonaqueous liquid systems for lithium-air batteries are electrolytes based on sulfones, ethers, carbonates, and amides, which can bear a broader electrochemical window than aqueous systems [39–42]. Among the various liquid electrolyte systems, Li triflate (LiTF) in tetra-ethylene glycol dimethyl ether (TEGDME) exhibited preferable electrochemical performance and relatively high stability due to its high stability and low volatility, which was favorable for fabricating flexible lithium-air batteries [43,44]. For example, a series of flexible lithium-air batteries were fabricated based on this electrolyte, exhibiting high electrochemical stability [9]. Ionic liquids are low-temperature molten salts and are also used as new electrolytes for lithium-air batteries due to their negligible vapor pressure, superior hydrophobicity, low flammability, and broad electrochemical window [45]. However, ionic liquids have much higher viscosity and relatively lower O_2 diffusivity coefficients. Although nonaqueous liquid electrolytes have exhibited certain advantages for fabricating flexible lithium-air batteries, the toxic and flammable organic liquid electrolytes cause a significant safety hazard. In addition, the leakage problem will be even worse when batteries are subjected to frequent bending, twisting, and other deformations over a long period. Therefore, nonaqueous liquid electrolytes are not a preeminent candidate in developing flexible lithium-air batteries.

Solid electrolytes have been investigated for many years and represent a good solution for electrolyte leakage. In addition, the pursuit of solid electrolytes is also intended to broaden the working potential window, improve battery safety, and develop all solid-state batteries [46]. Based on their compositions, solid electrolytes are classified into inorganic and polymer solid electrolytes (including polymer gel electrolytes). Among them, inorganic solid electrolytes mainly include NASICON ($Li_{1+x}Al_xTi_{2-x}$ $(PO_4)_3$, $Li_{1+x}Al_yGe_{2-y}(PO_4)_3$, perovskite ($La_{2/3-x}Li_{3x}TiO_3$), inverse perovskite ($Li_3OX$, Li_2OHX (X=Cl, Br, F)), and GARNET ($Li_7La_3Zr_2O_{12}$) [47–50]. However, the solid electrolytes are rigid and are not suitable for assembling flexible lithium-air batteries; what is worse, the ionic conductivity of this inorganic solid electrolyte is also low. As for assembling flexible metal-air batteries, solid electrolytes with flexible properties must be considered, and the recent development of polymer solid electrolytes represents an excellent solution to this problem. Compared with the fragility and poor mechanical flexibility of inorganic solid electrolytes, polymer electrolytes with high flexibility, low interfacial impedance, and high ionic conductivity are promising alternatives to liquid electrolytes, especially under repeated bending conditions [51]. In consideration of its intrinsic flexibility and protection for active Li metal anode, polymer electrolytes are highly compatible with flexible lithium-air batteries. Polymer solid electrolytes mainly include polymers and Li salts, which play the roles of framework supports and ionic conductors, respectively. Among them, the polymer matrix mainly includes polyethylene oxide (PEO), polyvinylidene fluoride (PVDF), poly(vinylidene fluoride-*co*-hexafluoropropylene) (PVDF-HFP), and polyacrylonitrile [52,53]. As shown in Figure 7.5a, a flexible polymer solid electrolyte is prepared based on 3D Li-ion-conducting ceramic network to provide continuous Li-ion transfer channels in the PEO matrix, which exhibited an ionic conductivity of 2.5×10^{-4} S·cm^{-1} at room temperature [54]. The electrolyte could also effectively block dendrites in the lithium-ion battery. In addition, a series of hydrophobic and free-standing polymer gel electrolytes had been developed from PVDF-HFP and were successfully used for flexible lithium-air batteries (Figure 7.5b) [55].

7.2 FLEXIBLE THIN-FILM LITHIUM-AIR BATTERIES

Stable electrochemical performance under repetitive external force is crucial for flexible batteries. To this end, a novel battery configuration is required to enable high electrochemical and mechanical stabilities. At present, flexible lithium-air batteries are mainly made with planar and fiber structures, and these designs can satisfy mechanical deformations. Planar flexible lithium-air batteries have imitated the commonly researched lithium-ion batteries, and the only different point is the permeability of the packaging. Generally, the metal electrode is attached to a metal foil as a substrate/current collector to ensure good electrical contact. Then, an electrolyte membrane is sandwiched between the metal anode

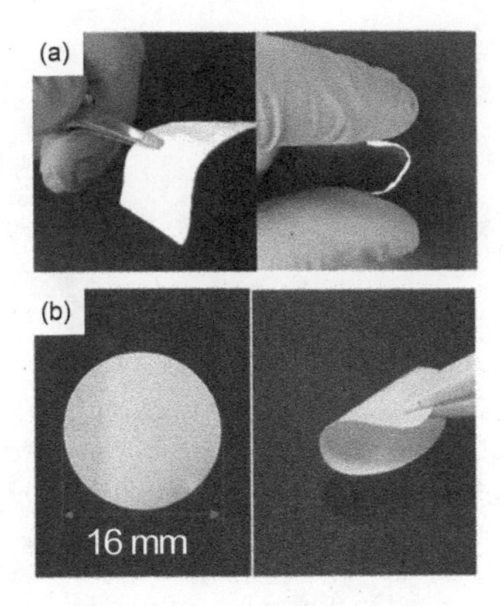

FIGURE 7.5 Photographs of the flexible and bendable solid electrolytes based on (a) polyethylene oxide (PEO) matrix and (b) poly(vinylidene fluoride-*co*-hexafluoropropylene) (PVDF-HFP) matrix. (Reproduced from Ref. [54] with permission of the National Academy of Sciences, USA, and Ref. [55] with permission of Wiley-VCH.)

and the air cathode. A porous current collector is attached to the air cathode for gas diffusion and electron transfer.

For example, a flexible thin-film lithium-air battery was built by assembling a flexible cathode based on $NiCo_2O_4$ nanowire array-coated carbon textile, a separator, and a Li foil anode (Figure 7.6a) [56]. The obtained lithium-air battery showed high flexibility, as presented in Figure 7.6b, and a green light-emitting diode (LED) can always be lit up when the thin-film lithium-air battery was bent into a circle. Thanks to its high bendable properties, the flexible thin-film lithium-air battery has potential applications in flexible and wearable electronic devices, such as roll-up displays, wearable sensors, and irregularly shaped energy-storage devices.

For example, a flexible thin-film lithium-air battery is fabricated by containing a flexible air cathode based on TiO_2 NA-modified carbon textile, a glass-fiber membrane, and a Li foil anode (Figure 7.7a) [9]. The as-fabricated lithium-air battery exhibited excellent electrochemical performance and flexibility. The battery could sustain a stable cycling life for more than 350 cycles with a capacity limit of 500 $mAh \cdot g^{-1}$ at a current density of 100 $mAh \cdot g^{-1}$ (Figure 7.7b). In order to verify its potential applications in the field of flexible electronics, severely bent and twisted batteries

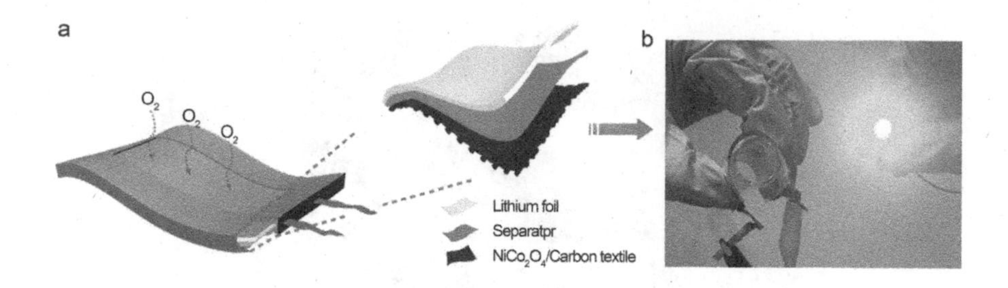

FIGURE 7.6 (a) Schematic illustration of the structure of the flexible thin-film lithium-air battery based on a Li foil anode and a $NiCo_2O_4$-coated carbon textile cathode. (b) The lithium-air battery powering a commercial green LED under bending. (Reproduced from Ref. [56] with permission of the American Chemical Society.)

at various angles were used to power a commercial red LED display screen (Figure 7.7c–e). Even after 100 cycles, the terminal discharge-charge voltage of the thin-film lithium-air battery remained almost constant, showing high flexibility and stability.

Interestingly, inspired by Chinese brush painting and writing, a flexible thin-film lithium-air battery was recently developed from ink/paper cathode, glass-fiber membrane, and Li foil anode (Figure 7.8a) [57]. The flexible ink/paper cathode was prepared by a facile and scalable fabrication method, using a brush to coat catalyst ink on commonly used paper. This low cost and facile synthesis method is quite attractive for industrial-scale production. As presented in Figure 7.8b, the assembled lithium-air battery can power commercial red LED equipment in both planar and bent conditions. Furthermore, the first discharge curves of the thin-film lithium-air battery before and after bending 1,000 cycles and the corresponding rate performance are almost the same, showing high flexibility and stability (Figure 7.8c and d). More interestingly, inspired by the unique characteristics of this ink/paper cathode, a foldable lithium-air battery pack consisting of several paralleled cells is prepared for the first time (Figure 7.8e). As a result, the unique foldable structure can play a vital role as a power switch, which enabled the lithium-air battery to provide power in a compressed state that automatically stopped after the external force was released (Figure 7.8f).

7.3 FIBER LITHIUM-AIR BATTERIES

Another configuration of flexible lithium-air battery is fabricated by designing a fiber shape. Compared to the sandwiched structure of thin-film

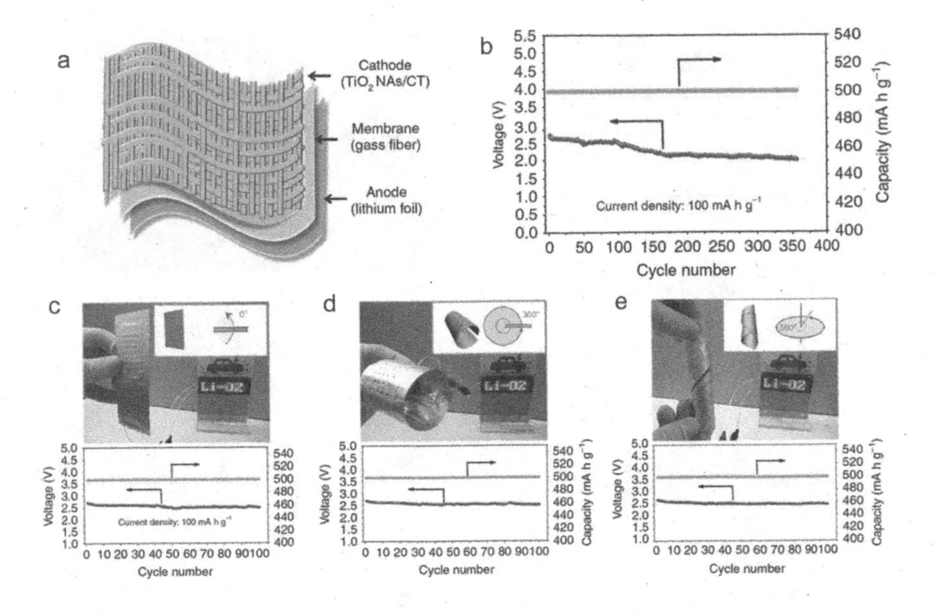

FIGURE 7.7 (a) Schematic illustration of the structure of the flexible thin-film lithium-air battery based on a Li foil anode and a TiO_2 NA-modified carbon textile cathode. (b) Cycling performance of the lithium-air battery. (c–e) Charge-discharge curves and the corresponding cycling performance of the lithium-air battery under pristine, bending, and twisting conditions, respectively. (Reproduced from Ref. [9] with permission of Nature Publishing Group.)

lithium-air batteries, the fiber structure attracts excellent attention in the field of flexible electronics due to its improved flexibility, which can be further woven into textile and applied in wearable electronics. Typically, it contains a Li wire or Li rod electrode in the center, wrapped by an electrolyte layer. The catalyst-loaded air electrode is then wound on the electrolyte layer, and these components are finally wrapped with a perforated packing insulator. As a typical example, an all-solid-state fiber lithium-air battery is developed based on a Li wire anode, a polymer gel electrolyte, and an aligned CNT sheet air cathode (Figure 7.9a and b) [58]. First, a layer of PVDF-HFP-based polymer gel electrolyte is coated on the Li wire. An aligned CNT sheet was then wrapped around to function as an air electrode, and a punched heat-shrinkable tube was finally packed around the whole assembly to prevent the lithium-air battery from being damaged. In this system, the gel electrolyte served not only as an ion conductor but also as an effective Li metal protector to prevent side reactions in air, and thus it can stably work in ambient air. The resulting lithium-air battery exhibited

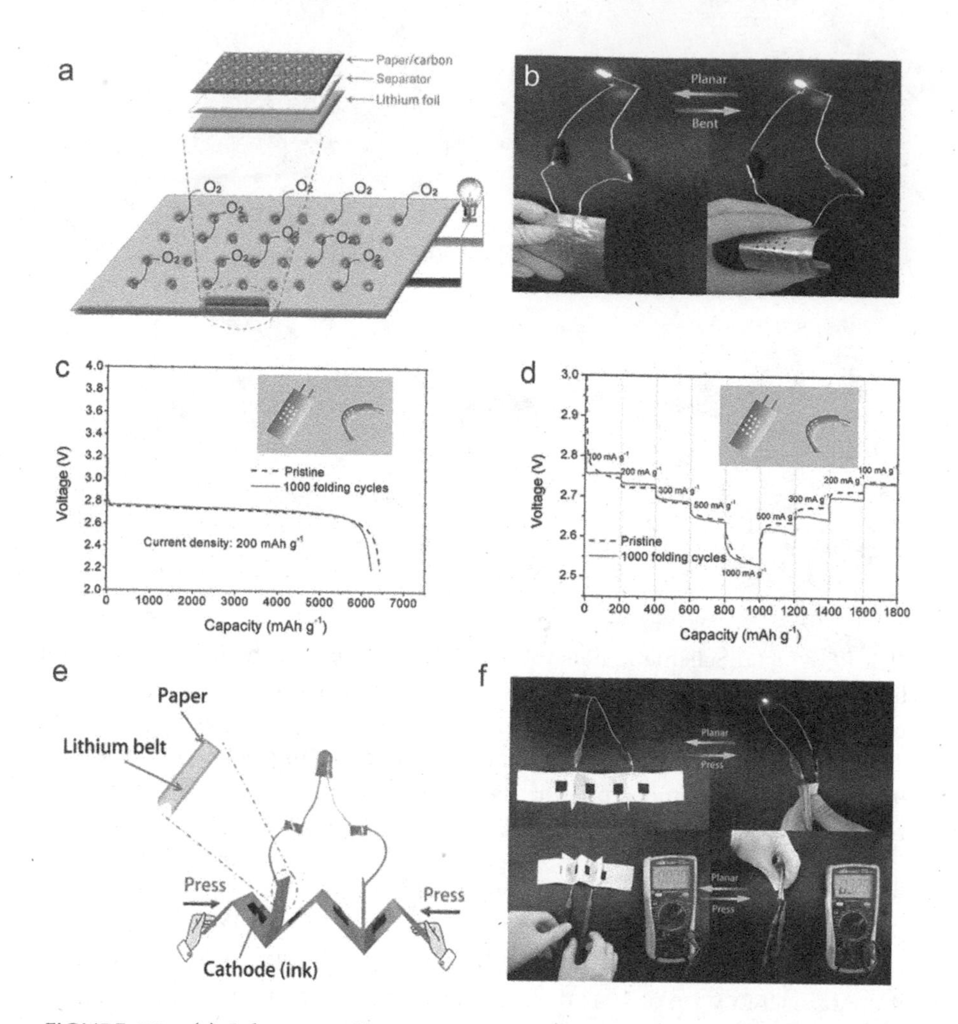

FIGURE 7.8 (a) Schematic illustration of the structure of the flexible thin-film lithium-air battery based on a Li foil anode and an ink/paper cathode. (b) The lithium-air battery powering a commercial red LED under planar and bending conditions. (c) First discharge curves of the lithium-air battery before and after bending for 1,000 cycles. (d) Rate capability of the lithium-air battery before and after bending for 1,000 cycles at different current densities. (e) Schematic illustration of the working mechanism of a foldable lithium-air battery pack. (f) Photographs of a lithium-air battery pack powering a commercial red LED. (Reproduced from Ref. [57] with permission of Wiley-VCH.)

a discharge capacity of 12,470 $mAh \cdot g^{-1}$ at a current density of 1,400 $mA \cdot g^{-1}$ and could effectively work for 100 cycles in the air with a cutoff capacity of 500 $mAh \cdot g^{-1}$. The fiber shape gave it high flexibility (Figure 7.9c), and the electrochemical performances were well maintained under and after bending. It could also be woven into flexible power textiles or knapsacks to support various electronic devices (Figure 7.9d and e).

Although the above fiber lithium-air battery exhibits high flexibility, superior electrochemical performances, and high mechanical stability, the low cost and scalable fabrication are also imperative characteristics needed to be considered for practical applications. To this end, a low cost air electrode was prepared by dyeing industrially weavable and highly conductive metal/cotton yarns with active material ink (RuO_2-coated nitrogen-doped carbon nanotube) (Figure 7.10a) [59]. When it was further assembled into fiber lithium-air batteries (the commercial Li strip as the anode), it exhibited a high discharge capacity of 1,981 $mAh \cdot g^{-1}$ at a current density of 320 $mA \cdot g^{-1}$. The obtained fiber lithium-air battery also showed high flexibility. As presented in Figure 7.10b, the LED brightness remained almost unchanged when the fiber lithium-air battery was bent to 120°, 180°, or even the knot condition. In addition, the battery can be

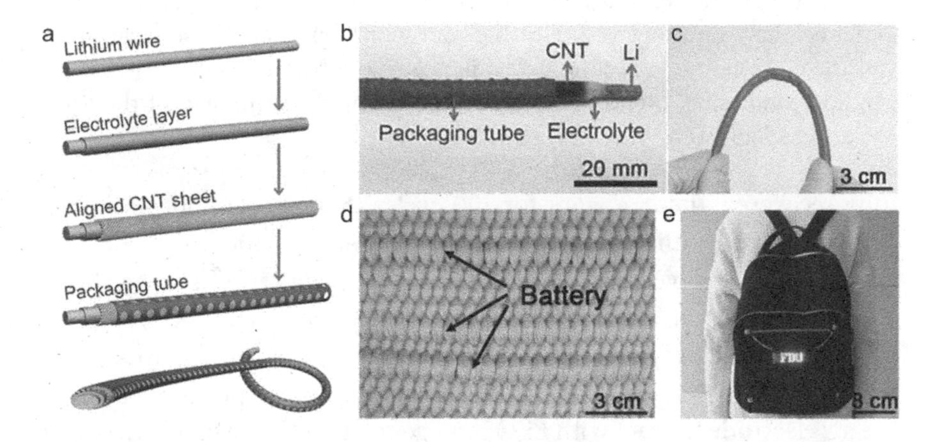

FIGURE 7.9 (a) Schematic illustration of the fabrication of the fiber lithium-air battery based on a Li wire anode and aligned carbon nanotube (CNT) sheet cathode. (b and c) Photographs of a fiber lithium-air battery and its bending condition, respectively. (d) Photograph of a lithium-air battery textile. (e) Photograph of a fiber lithium-air battery woven into a knapsack to power a commercial blue LED display screen. (Reproduced from Ref. [58] with permission of Wiley-VCH.)

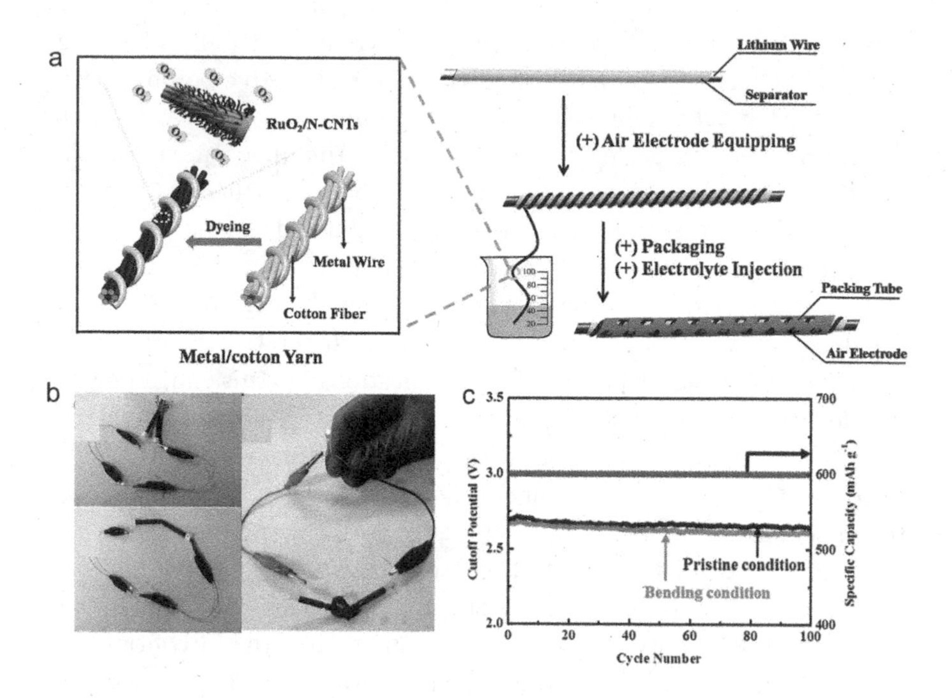

FIGURE 7.10 (a) Schematic illustration of the fabrication of the fiber lithium-air battery based on a Li wire anode and a highly conductive metal/cotton yarn cathode. (b) The lithium-air battery powering a commercial red LED equipment under bending, knotting, and releasing conditions. (c) Charge-discharge curves and the corresponding cycling performance of the battery under pristine and bending conditions. (Reproduced from Ref. [59] with permission of the Royal Society of Chemistry.)

fully recovered and can work for 100 cycles (>600 h) without noticeable degradation even under bending conditions (Figure 7.10c).

The ultimate development goal of lithium-air batteries is to be able to use O_2 directly from ambient air. Unfortunately, most of the work reported is based on pure oxygen or dry air atmosphere. When the lithium-air battery works in ambient air, water molecules in the air can permeate through the air electrode to react with Li_2O_2 and generate LiOH, which reacts with CO_2 to form Li_2CO_3. Li_2CO_3 is more chemically stable than Li_2O_2, which gradually accumulates to accelerate the failure of the lithium-air battery. Attempts have been made to suppress the water molecules from the ambient air by introducing another selective breathable layer around the battery. For instance, a flexible fiber lithium-air battery with ultralong cycle life in ambient air is reported by combining a low-density polyethylene

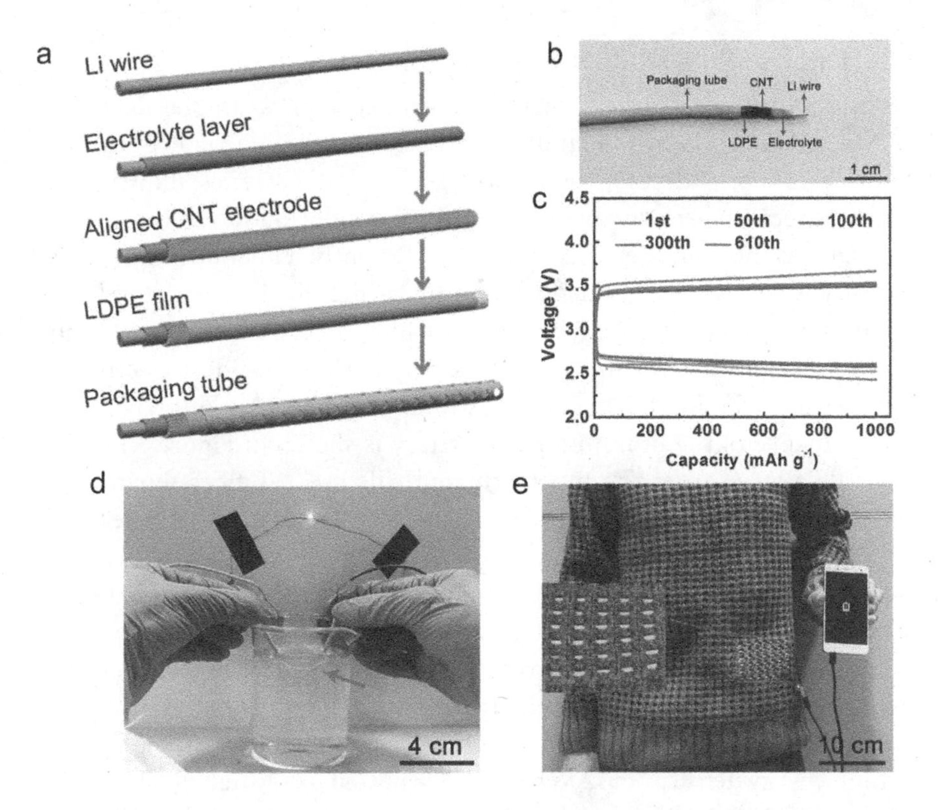

FIGURE 7.11 (a) Schematic illustration of the fabrication of the fiber lithium-air battery based on an low-density polyethylene (LDPE) film as a selective breathable layer. (b) Photograph of the fiber lithium-air battery. (c) Discharge-charge curves of a fiber lithium-air battery at a current density of 2,000 $mA \cdot g^{-1}$ in ambient air. (d) A fiber lithium-air battery being immersed in water to power a commercial yellow LED. (e) Photographs of fiber lithium-air batteries being woven into clothes to charge a smartphone. (Reproduced from Ref. [6] with permission of Wiley-VCH.)

(LDPE) film and a polymer gel electrolyte with LiI as a redox mediator (Figure 7.11a and b) [6]. The prepared fiber lithium-air battery exhibited a good cyclability that can discharge and charge for more than 300 cycles in ambient air (Figure 7.11c). Moreover, with the help of the waterproof LDPE film, this fiber lithium-air battery could stably power a yellow LED even when it was partially immersed in water (Figure 7.11d). Besides, the fiber lithium-air battery integrated into clothes can effectively charge a smartphone, which offered the possibility to manufacture wearable energy devices (Figure 7.11e).

7.4 FIBER LITHIUM-ION-AIR BATTERIES

To meet the rapid development of portable and wearable electronic devices, it is necessary to manufacture batteries with high energy density and flexibility. Although the flexible lithium-air batteries shown good electrochemical performance and appropriate deformation ability, the use of Li metal anode limited their flexibility. Recently, a fiber lithium-ion-air battery with high energy density and ultra-high flexibility is developed by designing a coaxial architecture with a lithiated silicon/carbon nanotube hybrid fiber as the inner anode, a polymer gel as middle electrolyte, and a bare CNT sheet as the outer cathode (Figure 7.12a) [60]. The working principle of the lithium-ion-air battery is shown in Figure 7.12b. Li^+ ions de-alloyed from the inner lithiated silicon/CNT fiber during discharge and then transferred through the gel electrolyte. O_2 diffused into the voids of aligned CNT sheets from all directions and reacted with Li^+ ions to form Li_2O_2 at the CNT cathode while the electrons flowed from the anode to the cathode through the external circuit. The process was reversed during charge. The coaxial structure also facilitated the diffusion of ions and oxygen molecules in all directions while maintaining the integrality during deformation. The fiber lithium-ion-air battery showed a high energy density of 512 $Wh{\cdot}kg^{-1}$ and showed ultra-high flexibility. As shown in Figure 7.12c, the fiber lithium-ion-air battery could withstand various deformations, including bending, tying, twisting, and looping, without affecting its charging and discharging behaviors (Figure 7.12d). The battery could effectively work even after 20,000 bending cycles (Figure 7.12e).

7.5 PERSPECTIVE

Flexible lithium-air batteries show a much higher theoretical energy density than other conventional flexible energy-storage systems. Therefore, it has achieved rapid development in the past few years and has become a promising candidate for an energy-storage device of flexible/wearable electronic products. However, the development of flexible lithium-air batteries is still in its infancy. Although some advanced battery models have been manufactured in recent years, there are still many challenges.

Compared with other batteries, the cycle life of the lithium-air batteries needs further development. This is mainly caused by the low catalyst activity and corrosion of the air environment (H_2O, N_2, CO_2, etc.).

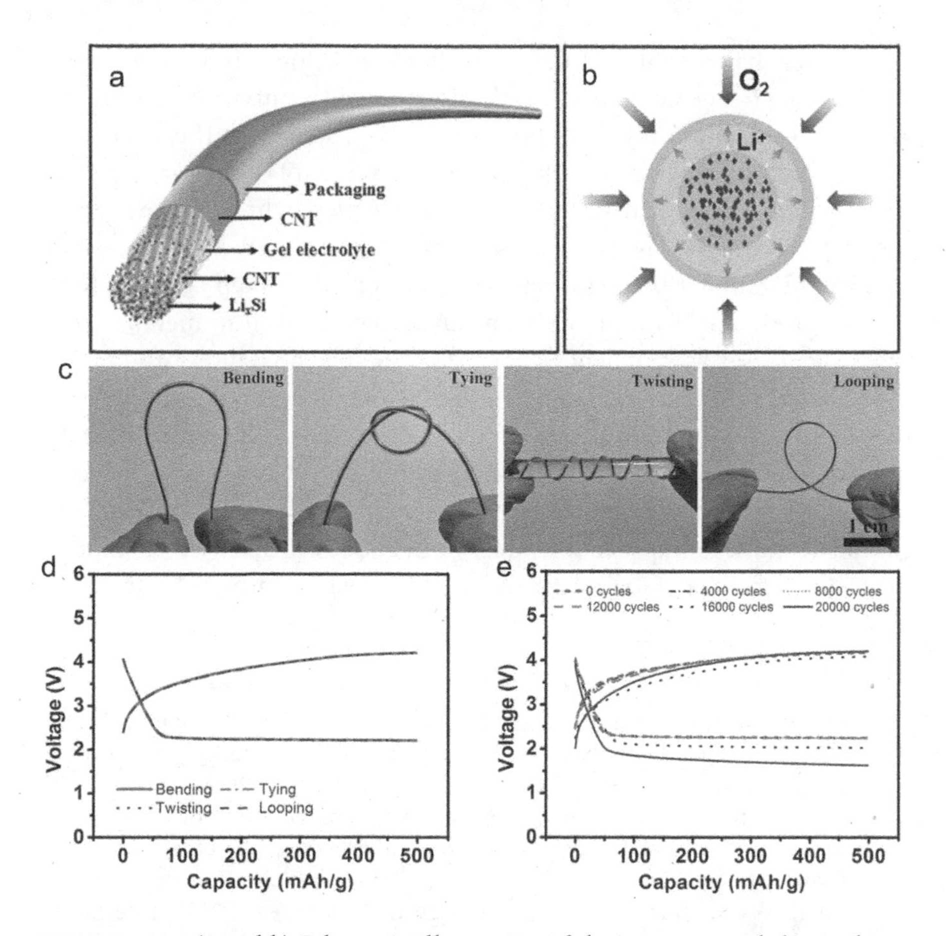

FIGURE 7.12 (a and b) Schematic illustration of the structure and the working mechanism of fiber lithium-ion-air battery, respectively. (c and d) Photographs and charge-discharge curves of the lithium-ion-air battery under various deformations. (e) Charge-discharge curves of the lithium-ion-air battery after different bending cycles. (Reproduced from Ref. [60] with permission of Wiley-VCH.)

Furthermore, the flexible lithium-air battery shows desirable flexibility but, in the meanwhile, erects obstacles for performance enhancement. Generally, the energy density of lithium-air batteries is very low, far below their theoretical value, because the mass proportion of the active material is relatively low. The active materials are easily peeled off the electrode during the repeated deformation processes, thus limiting the mass loading of active materials and deactivating the utilization of electrodes. To this end, maintaining a good balance between battery flexibility and electrochemical performance is worthy of attention.

Another challenge at the forefront of flexible lithium-air batteries is the safety issue. In the current research of flexible lithium-air batteries, the anodes are usually Li foils, Li ribbons, and Li sticks/wires. These metallic Li materials become one of the safety vulnerabilities of the battery because the inevitable Li dendrite formation is likely to pierce through the separator or gel electrolyte layer and cause a short circuit or even an explosion. In particular, under the deformation conditions that often occur in flexible batteries, cracks may appear on the surface of lithium metal, which intensifies the growth of dendrite and deteriorates the stability and safety of the battery. To this end, many studies have been devoted to improving the safety of Li anodes, such as applying artificial SEI layers, solid-state electrolytes, or replacing the Li metal with other Li-containing materials. Compared with the intensive research on air cathodes, more efforts should be made to improve the safety of Li anodes. Otherwise, Li anodes may soon become the main bottleneck of flexible lithium-air batteries.

REFERENCES

1. Tan, P. Chen, B. Xu, H. Zhang, H. Cai, W. Ni, M. Liu, M. Shao, Z. 2017. Flexible Zn-and Li-air batteries: recent advances, challenges, and future perspectives. *Energy & Environmental Science* 10: 2056–2080.
2. Zhang, S. S. Foster, D. Read, J. 2010. Discharge characteristic of a non-aqueous electrolyte Li/O_2 battery. *Journal of Power Sources* 195: 1235–1240.
3. Zhai, D. Lau, K. C. Wang, H. H. Wen, J. Miller, D. J. Lu, J. Kang, F. Li, B. Yang, W. Gao, J. Indacochea, E. Curtiss, L. A. Amine, K. 2015. Interfacial effects on lithium superoxide disproportionation in Li-O_2 batteries. *Nano Letters* 15: 1041–1046.
4. Lu, J. Cheng, L. Lau, K. C. Tyo, E. Luo, X. Wen, J. Miller, D. Assary, R. S. Wang, H.-H. Redfern, P. Wu, H. Park, J. B. Sun, Y. K. Vajda, S. Amine, K. Curtiss, L. A. 2014. Effect of the size-selective silver clusters on lithium peroxide morphology in lithium-oxygen batteries. *Nature Communications* 5: 4895.
5. Luntz, A. C. McCloskey, B. D. 2014. Nonaqueous Li-air batteries: a status report. *Chemical Reviews* 114: 11721–11750.
6. Wang, L. Pan, J. Zhang, Y. Cheng, X. Liu, L. Peng, H. 2018. A Li-air battery with ultralong cycle life in ambient air. *Advanced Materials* 30: 1704378.
7. Suen, N. Hung, S. Quan, Q. Zhang, N. Xu, Y. Chen, H. 2017. Electrocatalysis for the oxygen evolution reaction: recent development and future perspectives. *Chemical Society Reviews* 46: 337–365.
8. Wang, Z. Xu, D. Xu, J. Zhang, X. 2014. Oxygen electrocatalysts in metal-air batteries: from aqueous to non-aqueous electrolytes. *Chemical Society Reviews* 43: 7746–7786.

9. Liu, Q. Xu, J. Xu, D. Zhang, X. 2015. Flexible lithium-oxygen battery based on a recoverable cathode. *Nature Communications* 6: 7892.
10. Liu, Q. Xu, J. Chang, Z. Xu, D. Yin, Y. Yang, X. Liu, T. Jiang, Y. Yan, J. Zhang, X. 2016. Growth of Ru-modified Co_3O_4 nanosheets on carbon textiles toward flexible and efficient cathodes for flexible Li-O_2 batteries. *Particle & Particle Systems Characterization* 33: 500–505.
11. Wang, X. Li, W. Chen, Z. Waje, M. Yan, Y. 2006. Durability investigation of carbon nanotube as catalyst support for proton exchange membrane fuel cell. *Journal of Power Sources* 158: 154–159.
12. Itkis, D. M. Semenenko, D. A. Kataev, E. Y. Belova, A. I. Neudachina, V. S. Sirotina, A. P. Hävecker, M. Teschner, D. Knop-Gericke, A. Dudin, P. Barinov, A. Goodilin, E. A. Shao-Horn, Y. Yashina, L. V. 2013. Reactivity of carbon in lithium-oxygen battery positive electrodes. *Nano Letters* 13: 4697–4701.
13. McCloskey, B. D. Speidel, A. Scheffler, R. Miller, D. C. Viswanathan, V. Hummelshøj, J. S. Nørskov, J. K. Luntz, A. C. 2012. Twin problems of interfacial carbonate formation in non-aqueous Li-O_2 batteries. *The Journal of Physical Chemistry Letters* 3: 997–1001.
14. Ottakam Thotiyl, M. M. Freunberger, S. A. Peng, Z. Bruce, P. G. 2013. The carbon electrode in non-aqueous Li-O_2 Cells. *Journal of the American Chemical Society* 135: 494–500.
15. Cao, R. Lee, J. S. Liu, M. Cho, J. 2012. Recent progress in non-precious catalysts for metal-air batteries. *Advanced Energy Materials* 2: 816–829.
16. Leng, L. Zeng, X. Song, H. Shu, T. Wang, H. Liao, S. 2015. Pd nanoparticles decorating flower-like Co_3O_4 nanowire clusters to form an efficient, carbon/binder-free cathode for Li-O_2 batteries. *Journal of Materials Chemistry A* 3: 15626–15632.
17. Riaz, A. Jung, K. N. Chang, W. Lee, S. B. Lim, T. H. Park, S. J. Song, R. H. Yoon, S. Shin, K. H. Lee, J. W. 2013. Carbon-free cobalt oxide cathodes with tunable nanoarchitectures for rechargeable lithium-oxygen batteries. *Chemical Communications* 49: 5984–5986.
18. Kim, S. T. Choi, N. S. Park, S. Cho, J. 2015. Optimization of carbon- and binder-free Au nanoparticle-coated Ni nanowire electrodes for lithium-oxygen batteries. *Advanced Energy Materials* 5: 1401030.
19. Peng, Z. Freunberger, S. A. Chen, Y. Bruce, P. G. 2012. A reversible and higher-rate Li-O_2 Battery. *Science* 337: 563.
20. Zhao, Y. Yang, L. Chen, S. Wang, X. Ma, Y. Wu, Q. Jiang, Y. Qian, W. Hu, Z. 2013. Can boron and nitrogen Co-doping improve oxygen reduction reaction activity of carbon nanotubes? *Journal of the American Chemical Society* 135: 1201–1204.
21. Gong, K. Du, F. Xia, Z. Durstock, M. Dai, L. 2009. Nitrogen-doped carbon nanotube arrays with high electrocatalytic activity for oxygen reduction. *Science* 323: 760.
22. Wang, S. Zhang, L. Xia, Z. Roy, A. Chang, D. W. Baek, J.-B. Dai, L. 2012. BCN graphene as efficient metal-free electrocatalyst for the oxygen reduction reaction. *Angewandte Chemie International Edition* 51: 4209–4212.

23. Chen, P. Xiao, T. Y. Qian, Y. H. Li, S. S. Yu, S. H. 2013. A nitrogen-doped graphene/carbon nanotube nanocomposite with synergistically enhanced electrochemical activity. *Advanced Materials* 25: 3192–3196.
24. Liang, J. Jiao, Y. Jaroniec, M. Qiao, S. Z. 2012. Sulfur and nitrogen dual-doped mesoporous graphene electrocatalyst for oxygen reduction with synergistically enhanced performance. *Angewandte Chemie International Edition* 51: 11496–11500.
25. Chang, Z. Xu, J. Zhang, X. 2017. Recent progress in electrocatalyst for Li-O_2 batteries. *Advanced Energy Materials* 7: 1700875.
26. Jung, C. Y. Zhao, T. S. Zeng, L. Tan, P. 2016. Vertically aligned carbon nanotube-ruthenium dioxide core-shell cathode for non-aqueous lithium-oxygen batteries. *Journal of Power Sources* 331: 82–90.
27. Lu, X. Yin, Y. Zhang, L. Xi, L. Oswald, S. Deng, J. Schmidt, O. G. 2016. Hierarchically porous Pd/NiO nanomembranes as cathode catalysts in Li-O_2 batteries. *Nano Energy* 30: 69–76.
28. Liu, S. Zhu, Y. Xie, J. Huo, Y. Yang, H. Y. Zhu, T. Cao, G. Zhao, X. Zhang, S. 2014. Direct growth of flower-like δ-MnO_2 on three-dimensional graphene for high-performance rechargeable Li-O_2 batteries. *Advanced Energy Materials* 4: 1301960.
29. Lu, Y.-C. Xu, Z. Gasteiger, H. A. Chen, S. Hamad-Schifferli, K. Shao-Horn, Y. 2010. Platinum-gold nanoparticles: a highly active bifunctional electrocatalyst for rechargeable lithium–air batteries. *Journal of the American Chemical Society* 132: 12170–12171.
30. Li, P. Sun, W. Yu, Q. Guan, M. Qiao, J. Wang, Z. Rooney, D. Sun, K. 2015. An effective three-dimensional ordered mesoporous $ZnCo_2O_4$ as electrocatalyst for Li-O_2 batteries. *Materials Letters* 158: 84–87.
31. Lim, H.-D. Lee, B. Zheng, Y. Hong, J. Kim, J. Gwon, H. Ko, Y. Lee, M. Cho, K. Kang, K. 2016. Rational design of redox mediators for advanced Li-O_2 batteries. *Nature Energy* 1: 16066.
32. Chen, Y. Freunberger, S. A. Peng, Z. Fontaine, O. Bruce, P. G. 2013. Charging a Li-O_2 battery using a redox mediator. *Nature Chemistry* 5: 489–494.
33. Lim, H. D. Song, H. Kim, J. Gwon, H. Bae, Y. Park, K. Y. Hong, J. Kim, H. Kim, T. Kim, Y. H. Lepró, X. Ovalle-Robles, R. Baughman, R. H. Kang, K. 2014. Superior rechargeability and efficiency of lithium–oxygen batteries: hierarchical air electrode architecture combined with a soluble catalyst. *Angewandte Chemie International Edition* 53: 3926–3931.
34. Bergner, B. J. Schürmann, A. Peppler, K. Garsuch, A. Janek, J. 2014. TEMPO: a mobile catalyst for rechargeable Li-O_2 batteries. *Journal of the American Chemical Society* 136: 15054–15064.
35. Bi, X. Amine, K. Lu, J. 2020. The importance of anode protection towards lithium oxygen batteries. *Journal of Materials Chemistry A* 8: 3563–3573.
36. Wang, X. Pan, Z. Yang, J. Lyu, Z. Zhong, Y. Zhou, G. Qiu, Y. Zhang, Y. Wang, J. Li, W. 2019. Stretchable fiber lithium metal anode. *Energy Storage Materials* 22: 179–184.

37. Wang, L. Zhang, Y. Pan, J. Peng, H. 2016. Stretchable lithium-air batteries for wearable electronics. *Journal of Materials Chemistry A* 4: 13419–13424.
38. Li, F. Zhang, T. Zhou, H. 2013. Challenges of non-aqueous Li-O_2 batteries: electrolytes, catalysts, and anodes. *Energy & Environmental Science* 6: 1125–1141.
39. Balaish, M. Kraytsberg, A. Ein-Eli, Y. 2014. A critical review on lithium-air battery electrolytes. *Physical Chemistry Chemical Physics* 16: 2801–2822.
40. Walker, W. Giordani, V. Uddin, J. Bryantsev, V. S. Chase, G. V. Addison, D. 2013. A rechargeable Li-O_2 battery using a lithium nitrate/n, n-dimethylacetamide electrolyte. *Journal of the American Chemical Society* 135: 2076–2079.
41. Jung, H. G. Hassoun, J. Park, J. B. Sun, Y. K. Scrosati, B. 2012. An improved high-performance lithium-air battery. *Nature Chemistry* 4: 579–585.
42. Chen, Y. Freunberger, S. A. Peng, Z. Bardé, F. Bruce, P. G. 2012. Li-O_2 battery with a dimethylformamide electrolyte. *Journal of the American Chemical Society* 134: 7952–7957.
43. Adams, B. D. Radtke, C. Black, R. Trudeau, M. L. Zaghib, K. Nazar, L. F. 2013. Current density dependence of peroxide formation in the Li-O_2 battery and its effect on charge. *Energy & Environmental Science* 6: 1772–1778.
44. Freunberger, S. A. Chen, Y. Drewett, N. E. Hardwick, L. J. Bardé, F. Bruce, P. G. 2011. The lithium-oxygen battery with ether-based electrolytes. *Angewandte Chemie International Edition* 50: 8609–8613.
45. Guo, H. Luo, W. Chen, J. Chou, S. Liu, H. Wang, J. 2018. Review of electrolytes in non-aqueous lithium-oxygen batteries. *Advanced Sustainable Systems* 2: 1700183.
46. Choi, K. H. Kim, S. H. Ha, H. J. Kil, E. H. Lee, C. K. Lee, S. B. Shim, J. K. Lee, S. Y. 2013. Compliant polymer network-mediated fabrication of a bendable plastic crystal polymer electrolyte for flexible lithium-ion batteries. *Journal of Materials Chemistry A* 1: 5224–5231.
47. Thangadurai, V. Weppner, W. 2006. Recent progress in solid oxide and lithium ion conducting electrolytes research. *Ionics* 12: 81–92.
48. Arbi, K. Rojo, J. M. Sanz, J. 2007. Lithium mobility in titanium based Nasicon $Li_{1+x}Ti_{2-x}Al_x(PO_4)_3$ and $LiTi_{2-x}$ $Zr_x(PO_4)_3$ materials followed by NMR and impedance spectroscopy. *Journal of the European Ceramic Society* 27: 4215–4218.
49. Zhu, Y. Chu, W. Weppner, W. 2009. Investigation of gas concentration cell based on LiSiPO electrolyte and Li_2CO_3, Au electrode. *Chinese Science Bulletin* 54: 1334–1339.
50. van Wüllen, L. Echelmeyer, T. Meyer, H.-W. Wilmer, D. 2007. The mechanism of Li-ion transport in the garnet $Li_5La_3Nb_2O_{12}$. *Physical Chemistry Chemical Physics* 9: 3298–3303.
51. Manthiram, A. Yu, X. Wang, S. 2017. Lithium battery chemistries enabled by solid-state electrolytes. *Nature Reviews Materials* 2: 16103.

52. Vassal, N. Salmon, E. Fauvarque, J. F. 2000. Electrochemical properties of an alkaline solid polymer electrolyte based on P(ECH-co-EO). *Electrochimica Acta* 45: 1527–1532.
53. Thakur, V. K. Ding, G. Ma, J. Lee, P. S. Lu, X. 2012. Hybrid materials and polymer electrolytes for electrochromic device applications. *Advanced Materials* 24: 4071–4096.
54. Fu, K. Gong, Y. Dai, J. Gong, A. Han, X. Yao, Y. Wang, C. Wang, Y. Chen, Y. Yan, C. Li, Y. Wachsman, E. D. Hu, L. 2016. Flexible, solid-state, ion-conducting membrane with 3D garnet nanofiber networks for lithium batteries. *Proceedings of the National Academy of Sciences* 113: 7094.
55. Liu, T. Liu, Q. Xu, J. Zhang, X. 2016. Cable-type water-survivable flexible Li-O_2 battery. *Small* 12: 3101–3105.
56. Xue, H. Wu, S. Tang, J. Gong, H. He, P. He, J. Zhou, H. 2016. Hierarchical porous nickel cobaltate nanoneedle arrays as flexible carbon-protected cathodes for high-performance lithium-oxygen batteries. *ACS Applied Materials & Interfaces* 8: 8427–8435.
57. Liu, Q. Li, L. Xu, J. Chang, Z. Xu, D. Yin, Y. Yang, X. Liu, T. Jiang, Y. Yan, J. Zhang, X. 2015. Flexible and foldable Li-O_2 battery based on paper-ink cathode. *Advanced Materials* 27: 8095–8101.
58. Zhang, Y. Wang, L. Guo, Z. Xu, Y. Wang, Y. Peng, H. 2016. High-performance lithium–air battery with a coaxial-fiber architecture. *Angewandte Chemie International Edition* 55: 4487–4491.
59. Lin, X. Kang, Q. Zhang, Z. Liu, R. Li, Y. Huang, Z. Feng, X. Ma, Y. Huang, W. 2017. Industrially weavable metal/cotton yarn air electrodes for highly flexible and stable wire-shaped Li-O_2 batteries. *Journal of Materials Chemistry A* 5: 3638–3644.
60. Zhang, Y. Jiao, Y. Lu, L. Wang, L. Chen, T. Peng, H. 2017. An ultraflexible silicon-oxygen battery fiber with high energy density. *Angewandte Chemie International Edition* 56: 13741–13746.

CHAPTER 8

Flexible Aluminum-Air Batteries

8.1 OVERVIEW OF ALUMINUM-AIR BATTERIES

Metal-air batteries have been intensively investigated over the last decade due to their high energy density and capacity, relatively low cost, and stable discharge voltage. The theoretical energy densities of metal-air batteries are twice to ten times higher than those of lithium-ion batteries. Among different kinds of metal-air batteries, aluminum-air batteries promise future large-scale energy applications due to their low cost, high theoretical voltage, and theoretical energy density. For instance, the energy density is the second highest only to lithium and much higher than magnesium and zinc [1]. Compared with lithium-ion batteries, aluminum-air batteries have much lower cost and higher safety and recyclability due to the lower reactivity, easy handling, and excellent safety of aluminum.

8.1.1 Working Mechanism

A primary aluminum-air battery is usually composed of an aluminum anode, electrolyte, and air cathode (Figure 8.1). The commonly used electrolytes for aluminum-air batteries are aqueous solutions of sodium hydroxide (NaOH), potassium hydroxide (KOH), and sodium chloride (NaCl) [2]. The electrochemical reactions of aluminum-air batteries in alkaline electrolytes are as follows:

DOI: 10.1201/9781003273677-8

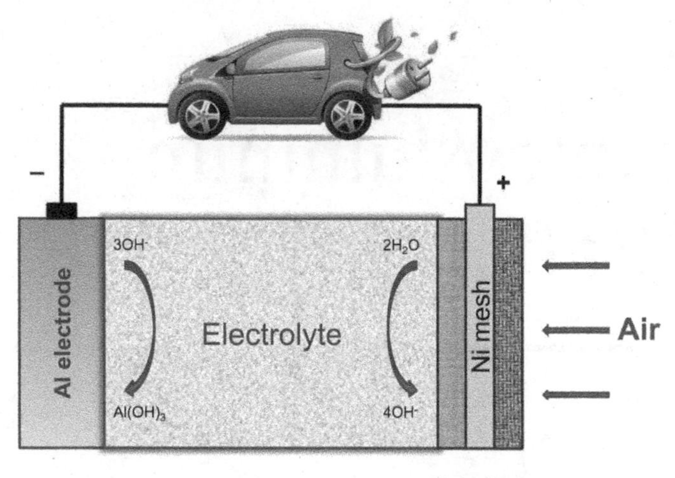

FIGURE 8.1 Schematic diagram of the structure of the aluminum-air battery. (Reproduced from Ref. [3] with permission of Elsevier.)

Anode:

$$Al \leftrightarrow Al^{3+} + 3e^-$$

Cathode:

$$O_2 + 2H_2O + 4e^- \leftrightarrow 4OH^-$$

Overall:

$$4Al + 3O_2 + 6H_2O \leftrightarrow 4Al(OH)_3$$

8.1.2 Electrode Active Materials

Due to its excellent electrochemical properties, pure metallic aluminum is generally used as anodic material for aluminum-air batteries, showing better anodic performances than aluminum anodes with impurities. The side reaction ($Al+3H_2O \rightarrow Al(OH)_3+3/2H_2$) may occur between the aluminum anode and water, which may cause corrosion and passivation of the aluminum anode surface and then lead to the failure of aluminum-air batteries. Therefore, many studies attempted to reduce the corrosion rate and hydrogen evolution reaction of aluminum-air batteries [4,5].

Although pure aluminum has many promising properties, it is unstable as anode for aluminum-air batteries. Aluminum alloys have thus been used as anode for aluminum-air batteries to overcome the limitations of pure aluminum by decreasing the corrosion rate and prolonging the battery operation time. Al-Zn, Al-Sn, Al-Ga, and Al-In alloys have been used as anode materials for aluminum-air batteries with relatively high performances [6,7]. Therefore, aluminum alloys can provide a decreased hydrogen evolution reaction and corrosion rate for aluminum-air batteries.

The air cathodes are critical components that will largely influence the electrochemical performances of aluminum-air batteries. The air cathodes usually comprise the gas diffusion layer, current collector, and catalytic active layer. The gas diffusion layer is generally composed of carbon material and a hydrophobic binder such as polytetrafluoroethylene to facilitate air diffusion and prevent the permeation of water. The current collector for aluminum-air batteries is typically made of metal nickel mesh or stainless steel connected to the external circuit to support the transfer processes of electrons [8,9]. The catalytic materials are usually mixed with binders on the current collectors as the catalytic active layer, where the main catalyzed chemical reaction of oxygen reduction reaction occurs [10,11].

Commercial aluminum-air batteries are hindered mainly by the sluggish efficiency of oxygen reduction reaction at the cathode. Therefore, it is critical to design appropriate cathodes with suitable catalytic materials to enhance the activity for oxygen reduction reaction for better aluminum-air batteries [3]. The overpotentials have limited the reaction kinetics of oxygen reduction reaction at the cathode, which can be enhanced by optimizing the catalytic materials [12]. Great efforts have been made to explore superior catalysts for cathodes of aluminum-air batteries, including precious metals and alloys [13,14], carbonaceous nanomaterials [15,16], transition metal oxides [17,18], carbon quantum dots [19], and metal-organic frameworks [20,21].

8.1.3 Electrolytes and Additives

The electrolyte is another essential component that influences the electrochemical performances of aluminum-air batteries, which controls the electrochemical reactions at the aluminum anode and determines the operating voltage of the batteries. The electrolytes for aluminum-air batteries can be divided into aqueous and nonaqueous electrolytes. For aqueous electrolytes, alkaline KOH and NaOH solutions with high ionic conductivity and low overpotential are the most used electrolytes. Aprotic electrolytes, solid-state electrolytes, and ionic liquids are usually used as nonaqueous electrolytes.

Based on the pH, aqueous electrolytes can be classified into alkaline electrolytes ($7<pH<13$), neutral salt electrolytes ($pH=7$), and acidic electrolytes ($2<pH<7$). In aqueous electrolytes, the self-corrosion reaction of aluminum and aluminum alloy anodes will cause the hydrogen evolution side reaction and loss of anode materials. In order to solve this problem, additives such as zinc oxide (ZnO) and sodium stannate (Na_2SnO_3) are used as corrosion inhibitors to suppress the undesired corrosion reactions by adsorbing on the surfaces of the aluminum anode [22,23].

8.1.4 Summary

Aluminum-air batteries are endowed with low cost, light weight, and high theoretical energy density. In addition, the abundant source of aluminum makes them one of the most promising and cost-effective power sources. Conventional aluminum-air batteries generally use rigid electrodes and liquid electrolytes, which have a rigid structure and cannot meet the flexible and lightweight requirements of portable and wearable electronics [24,25]. Therefore, the realization of flexibility, bendability, and portability for aluminum-air batteries is essential, especially for the advancement and application of flexible and wearable electronics.

8.2 FLEXIBLE THIN-FILM ALUMINUM-AIR BATTERY

Apart from improving the electrochemical performances, the design of flexible aluminum-air batteries with light weight, bendability, and portability is also essential, especially for flexible electronics. Compared with traditional aluminum-air batteries that are rigid and bulky, flexible aluminum-air batteries can normally work under bending, which are promising for the growing market of portable and wearable electronics. For instance, flexible thin-film aluminum-air batteries were fabricated by using a sodium polyacrylate (PANa)-based electrolyte (Figure 8.2a) with sodium carboxymethyl cellulose (CMC) prepared through free radical polymerization [26]. The synthesized PANa-CMC gel electrolyte showed a cobweb-like network microstructure. The uniformly distributed and interacted pores in the electrolyte displayed diameters of 10–20 μm with the coarse wall (Figure 8.2b and c), which could keep the channel stable and improve water-retention ability and

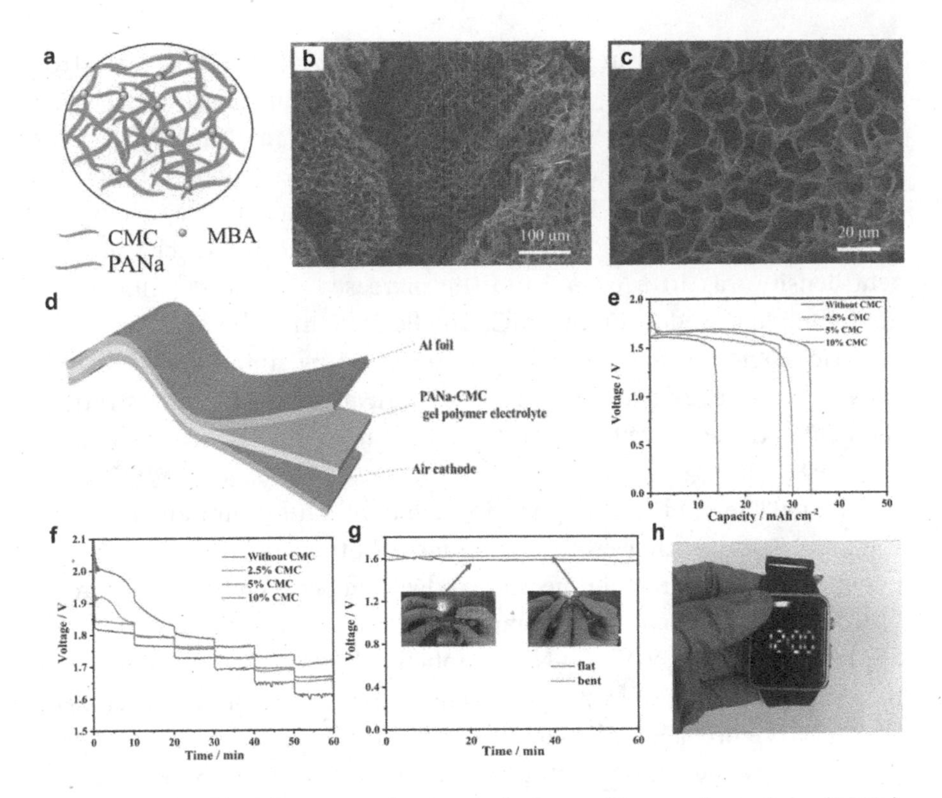

FIGURE 8.2 (a) Schematic diagram of the sodium polyacrylate (PANa)-carboxymethyl cellulose (CMC) gel polymer electrolyte. (b and c) SEM images of the PANa-5% CMC gel electrolyte at low and high magnifications, respectively. (d) Schematic illustration of the assembled flexible thin-film aluminum-air battery. (e) Galvanostatic discharge curves of the aluminum-air battery with different electrolytes at the current density of 1 mA cm^{-2}. (f) Rate performance of the aluminum-air battery at increasing current densities of 1, 2, 3, 5, 7.5, and 10 mA cm^{-2}. (g) Galvanostatic discharge curves of the thin-film aluminum-air battery with different electrolytes under bending at the current density of 3 mA cm^{-2}. (h) Application demonstration of the flexible thin-film aluminum-air batteries. (Reproduced from Ref. [26] with permission of IOP Publishing Ltd.)

ionic conductivity. The ionic conductivity of the used gel electrolyte (with PANa-5% CMC) was 0.324 S cm^{-1}. The thin-film aluminum-air batteries were fabricated by placing the PANa-CMC gel electrolyte between aluminum foil anode and MnO_2-decorated carbon paper air cathode with a sandwich-like structure (Figure 8.2d). Each part of the thin-film battery was flexible, which ensured good flexibility of the whole battery.

The thin-film aluminum-air battery with PANa-5% CMC electrolyte exhibited a discharge capacity of 34.5 mAh cm^{-2} and a discharge voltage of 1.86 V at 1 mA cm^{-2} (Figure 8.2e). The rate performance of the aluminum-air batteries with PANa-CMC gel electrolytes at increasing current densities of 1, 2, 3, 5, 7.5, and 10 mA cm^{-2} was shown in Figure 8.2f. The discharge voltage was 1.72 V when the discharge current density was 10 mA cm^{-2} and the increased discharge voltage was attributed to the addition of CMC. The flexible thin-film aluminum-air batteries could normally discharge under bending, and the bent battery showed no voltage decay (Figure 8.2g). As an application demonstration, the bent batteries were fixed on a watch strap to power the smartwatch (Figure 8.2h).

The flexible and reconfigurable shape of aluminum-air batteries (Figure 8.3a) was achieved through a facile bottom-down design [27]. A polymer electrolyte membrane was sandwiched between the air electrode (γ-MnO_2-N/S graphene as the catalyst) and aluminum foil (Figure 8.3b). SEM image of the γ-MnO_2-N/S graphene showed that the flaky and nanorod-like shapes of MnO_2 were densely distributed on the surfaces of graphene (Figure 8.3c). TEM images confirmed that the MnO_2 nanoparticles closely adhered to the graphene nanosheet (Figure 8.3d and e). The flexible aluminum-air battery exhibited a discharge capacity of ~1,200 mAh g^{-1} and a discharge voltage of ~1.4 V at the current density of 5 mA cm^{-2} (Figure 8.3f). The distribution of energy density and the specific capacity of the flexible aluminum-air battery at different current densities are shown in Figure 8.3g. The thin-film aluminum-air battery showed high flexibility and can be used as the energy source for flexible electronic devices (Figure 8.3h).

8.3 FIBER ALUMINUM-AIR BATTERIES

Besides thin-film aluminum-air batteries, fiber aluminum-air batteries are another kind of flexible aluminum-air batteries. One-dimensional fiber energy storage devices have attracted increasing interest in the past decade due to their unique advantages, such as being highly flexible, lightweight, and even stretchable, woven into textiles that can bear different deformations. A new family of all-solid-state fiber aluminum-air batteries was created (Figure 8.4a) by sequentially coating hydrogel electrolyte and cross-stacked carbon nanotube (CNT)/silver nanoparticle hybrid sheets air cathode onto a spring-like Al substrate [28]. The cross-linked hydrogel

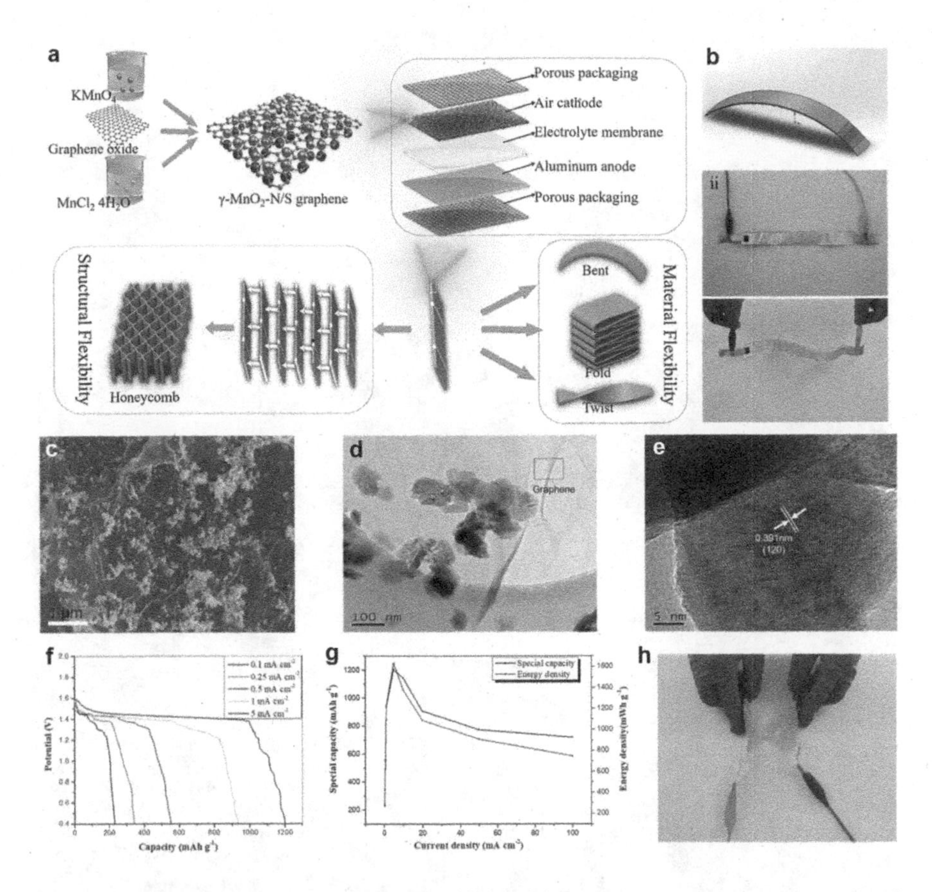

FIGURE 8.3 (a) Schematic diagram of the fabrication process, structure, and conceptual illustration of the flexible aluminum-air battery. (b) Conceptual illustration and photographs of the flexible aluminum-air battery. (c) SEM image of the γ-MnO_2-N/S graphene catalyst. (d and e) TEM and HRTEM images of the γ-MnO_2-N/S graphene catalyst, respectively. (f) Galvanostatic discharge curves of the aluminum-air battery at increasing current densities from 0.1 to 5 mA cm^{-2}. (g) Specific capacity and energy density of the aluminum-air battery at increasing current densities. (h) Photograph of the flexible thin-film aluminum-air battery under folding. (Reproduced from Ref. [27] with permission of the American Chemical Society.)

electrolyte was composed of poly(vinyl alcohol), poly(ethylene oxide), and KOH solution with additives of ZnO and Na_2SnO_3 to reduce the corrosion of the aluminum anode, which could be easily bent or stretched (Figure 8.4b and c) and showed a high ionic conductivity of 0.18 S cm^{-1} (Figure 8.4d).

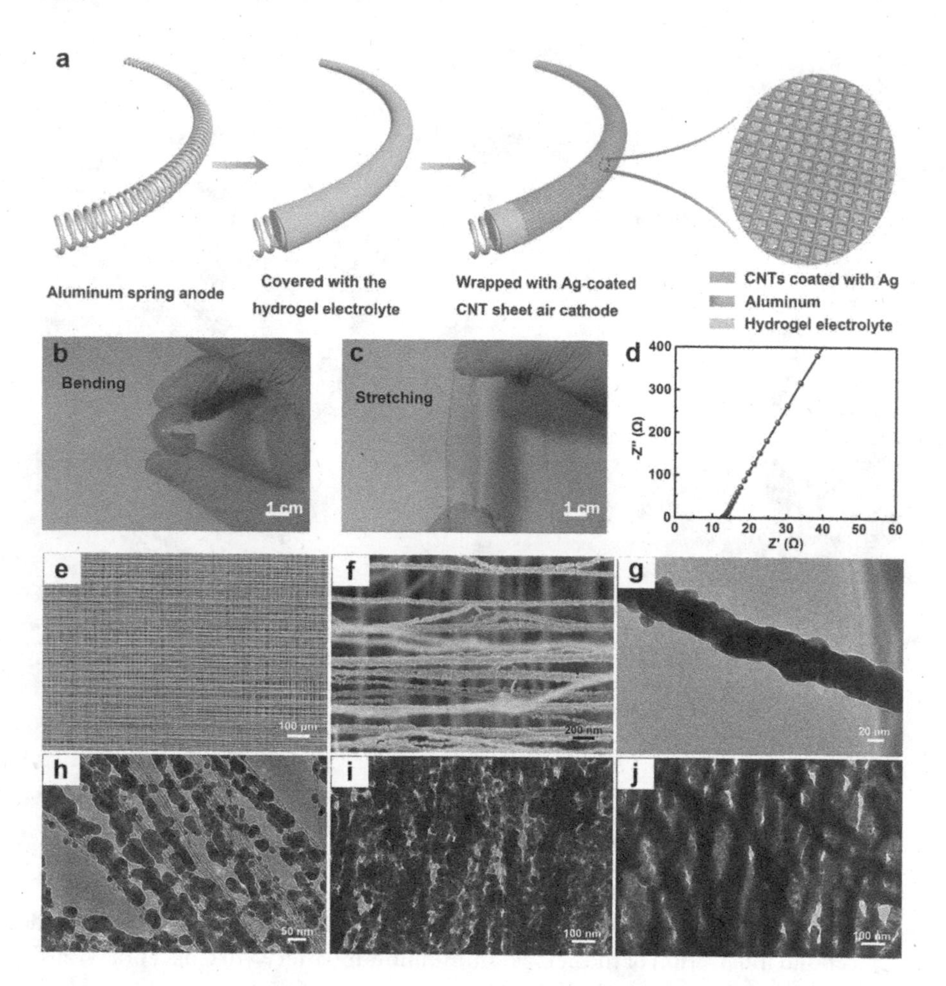

FIGURE 8.4 (a) Schematic diagram of the fabrication process of fiber aluminum-air battery. (b and c) Photographs of the cross-linked hydrogel electrolyte under bending and stretching, respectively. (d) Nyquist plot of the hydrogel electrolyte. (e and f) SEM images of the cross-stacked carbon nanotube (CNT)/silver nanoparticle hybrid sheets air cathode with a silver load of 45.5 μg cm^{-2} at low and high magnifications, respectively. (g) TEM image of the silver-coated CNT hybrid sheets air cathode at high magnification. (h–j) TEM images of the silver-coated CNT hybrid sheets air cathode with increasing silver load contents of 23.3, 45.5, and 66.9 μg cm^{-2}, respectively. (Reproduced from Ref. [28] with permission of Wiley-VCH.)

The CNT/silver nanoparticle hybrid sheets air cathode (with a silver load of 45.5 μg cm^{-2}) showed well-aligned and cross-stacked structure with the silver nanoparticles uniformly deposited on the surfaces of CNTs (Figure 8.4e and f) for effective oxygen reduction reaction sites, which was further proven by the high-resolution TEM image (Figure 8.4g). With the silver load content increased from 23.3, 45.5 to 66.9 μg cm^{-2}, the silver nanoparticles were well coated on the surfaces of aligned CNTs without apparent aggregation (Figure 8.4h–j). The fiber aluminum-air batteries with silver nanoparticles-coated air cathodes showed better discharge performances than bare CNT sheets at the same voltage (Figure 8.5a and b). The batteries with cross-stacked air cathode exhibited better rate discharge properties than those of other stacked patterns (Figure 8.5c).

The fiber aluminum-air battery (cross-stacked air cathode with a silver load of 45.5 μg cm^{-2}) displayed a specific capacity of 935 mAh g^{-1} with an energy density of 1,168 Wh kg^{-1} (Figure 8.5d). Under different bending angles, the output voltage of the fiber aluminum-air battery remained almost unchanged, showing good flexibility (Figure 8.5e). In addition, the fiber aluminum-air battery could withstand a stretching up to 30% with the output voltage maintained above 1 V (Figure 8.5f). The output voltage and power could be doubled by connecting two fiber aluminum-air batteries in series (Figure 8.5g). As an application demonstration, two series-connected fiber aluminum-air batteries were woven into a flexible textile to power a light-emitting diode watch worn on a human wrist (Figure 8.5h).

8.4 PERSPECTIVE

Aluminum-air batteries are promising candidates for future large-scale energy applications because of their low cost and high theoretical energy density. On the one hand, it is critical to design appropriate electrolytes and cathodes with suitable catalytic materials to enhance the oxygen reduction reaction activity. On the other hand, it is essential to develop flexible, bendable, and portable aluminum-air batteries to satisfy the fast-growing demands for next-generation flexible and wearable electronics. Therefore, thin-film and fiber aluminum-air batteries have been developed in recent years.

Despite the significant advances in flexible aluminum-air batteries, some challenges still exist to hinder their large-scale applications. First, the electrochemical performances (such as specific capacity and rate capability) of the flexible aluminum-air batteries need further improvements,

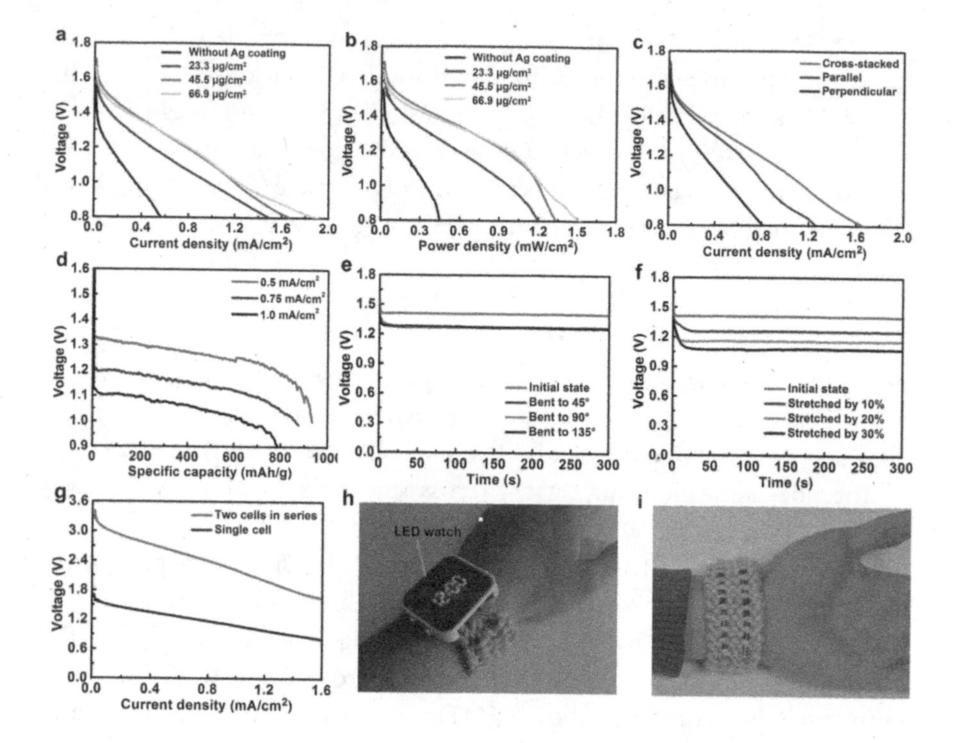

FIGURE 8.5 (a and b) Polarization and power density curves of the fiber aluminum-air batteries with different silver contents on the air cathode. (c) Polarization curves of the fiber aluminum-air batteries with different stacked patterns of the air cathodes. (d) Galvanostatic discharge curves of the fiber aluminum-air batteries at different current densities. (e and f) Galvanostatic charge-discharge curves of the fiber aluminum-air batteries under different bending angles and stretching ratios at a discharge current of 1 mA, respectively. (g) Polarization curves of a fiber aluminum-air battery and two batteries connected in series. (h and i) Application demonstration of the fiber-shaped aluminum-air batteries. Scale bar: 2 cm. (Reproduced from Ref. [28] by permission of Wiley-VCH.)

including the rational design of cathodes. Second, it is crucial to develop electrolytes with high ionic conductivity and low overpotential and suppress the undesirable corrosion reactions of aluminum. Third, an insightful understanding of the reaction mechanisms of flexible aluminum-air batteries during long-term working is also of great importance. Last, great efforts should be made to develop simple, controllable, and rapid preparation procedures for flexible aluminum-air batteries for their practical applications at a large scale.

REFERENCES

1. Mori, R. 2020. Recent developments for aluminum-air batteries. *Electrochemical Energy Reviews* 3: 344–369.
2. Sun, Z. Lu, H. Fan, L. Hong, Q. Leng, J. Chen, C. 2015. Performance of Al-air batteries based on Al-Ga, Al-In and Al-Sn alloy electrodes. *Journal of the Electrochemical Society* 162: A2116.
3. Liu, Y. Sun, Q. Li, W. Adair, K. R. Li, J. Sun, X. 2017. A comprehensive review on recent progress in aluminum-air batteries. *Green Energy & Environment* 2: 246–277.
4. Cho, Y.-J. Park, I.-J. Lee, H.-J. Kim, J.-G. 2015. Aluminum anode for aluminum-air battery-Part I: Influence of aluminum purity. *Journal of Power Sources* 277: 370–378.
5. Fan, L. Lu, H. Leng, J. 2015. Performance of fine structured aluminum anodes in neutral and alkaline electrolytes for Al-air batteries. *Electrochimica Acta* 165: 22–28.
6. El Shayeb, H. Abd El Wahab, F. El Abedin, S. Z. 2001. Electrochemical behaviour of Al, Al-Sn, Al-Zn and Al-Zn-Sn alloys in chloride solutions containing stannous ions. *Corrosion Science* 43: 655–669.
7. El Abedin, S. Z. Endres, F. 2004. Electrochemical behaviour of Al, Al-In and Al-Ga-In alloys in chloride solutions containing zinc Ions. *Journal of Applied Electrochemistry* 34: 1071–1080.
8. Sun, Z. Lu, H. 2015. Performance of Al-0.5 in as anode for Al-air battery in inhibited alkaline solutions. *Journal of the Electrochemical Society* 162: A1617.
9. Liu, K. Peng, Z. Wang, H. Ren, Y. Liu, D. Li, J. Tang, Y. Zhang, N. 2017. Fe_3C@Fe/N doped graphene-like carbon sheets as a highly efficient catalyst in Al-air batteries. *Journal of the Electrochemical Society* 164: F475.
10. Jung, K.-N. Kim, J. Yamauchi, Y. Park, M.-S. Lee, J.-W. Kim, J. H. 2016. Rechargeable lithium-air batteries: a perspective on the development of oxygen electrodes. *Journal of Materials Chemistry A* 4: 14050–14068.
11. Li, C. S. Sun, Y. Gebert, F. Chou, S. L. 2017. Current progress on rechargeable magnesium-air battery. *Advanced Energy Materials* 7: 1700869.
12. Goel, P. Dobhal, D. Sharma, R. 2020. Aluminum-air batteries: a viability review. *Journal of Energy Storage* 28: 101287.
13. Wang, Y.-J. Zhao, N. Fang, B. Li, H. Bi, X. T. Wang, H. 2015. Carbon-supported Pt-based alloy electrocatalysts for the oxygen reduction reaction in polymer electrolyte membrane fuel cells: particle size, shape, and composition manipulation and their impact to activity. *Chemical Reviews* 115: 3433–3467.
14. Wang, Z.-L. Xu, D. Xu, J.-J. Zhang, X.-B. 2014. Oxygen electrocatalysts in metal-air batteries: from aqueous to nonaqueous electrolytes. *Chemical Society Reviews* 43: 7746–7786.
15. Wu, G. Santandreu, A. Kellogg, W. Gupta, S. Ogoke, O. Zhang, H. Wang, H.-L. Dai, L. 2016. Carbon nanocomposite catalysts for oxygen reduction and evolution reactions: from nitrogen doping to transition-metal addition. *Nano Energy* 29: 83–110.

16. Dai, L. Xue, Y. Qu, L. Choi, H.-J. Baek, J.-B. 2015. Metal-free catalysts for oxygen reduction reaction. *Chemical Reviews* 115: 4823–4892.
17. Mainar, A. R. Colmenares, L. C. Leonet, O. Alcaide, F. Iruin, J. J. Weinberger, S. Hacker, V. Iruin, E. Urdanpilleta, I. Blazquez, J. A. 2016. Manganese oxide catalysts for secondary zinc air batteries: from electrocatalytic activity to bifunctional air electrode performance. *Electrochimica Acta* 217: 80–91.
18. Cordoba, M. Miranda, C. Lederhos, C. Coloma-Pascual, F. Ardila, A. Fuentes, G. A. Pouilloux, Y. Ramírez, A. 2017. Catalytic performance of Co_3O_4 on different activated carbon supports in the benzyl alcohol oxidation. *Catalysts* 7: 384.
19. Van Tam, T. Kang, S. G. Babu, K. F. Oh, E.-S. Lee, S. G. Choi, W. M. 2017. Synthesis of B-doped graphene quantum dots as a metal-free electrocatalyst for the oxygen reduction reaction. *Journal of Materials Chemistry A* 5: 10537–10543.
20. Furukawa, H. Cordova, K. E. O'Keeffe, M. Yaghi, O. M. 2013. The chemistry and applications of metal-organic frameworks. *Science* 341.
21. Gonen, S. Lori, O. Cohen-Taguri, G. Elbaz, L. 2018. Metal organic frameworks as a catalyst for oxygen reduction: an unexpected outcome of a highly active Mn-MOF-based catalyst incorporated in activated carbon. *Nanoscale* 10: 9634–9641.
22. Jafarian, M. Olyaei, H. M. B. Gobal, F. Hosseini, S. Mahjani, M. 2013. Study of the alloying additives and alkaline zincate solution effects on the commercial aluminum as galvanic anode for use in alkaline batteries. *Materials Chemistry and Physics* 143: 133–142.
23. Wang, X. Wang, J. Wang, Q. Shao, H. Zhang, J. 2011. The effects of polyethylene glycol (PEG) as an electrolyte additive on the corrosion behavior and electrochemical performances of pure aluminum in an alkaline zincate solution. *Materials and Corrosion* 62: 1149–1152.
24. Gelman, D. Shvartsev, B. Ein-Eli, Y. 2014. Aluminum-air battery based on an ionic liquid electrolyte. *Journal of Materials Chemistry A* 2: 20237–20242.
25. Wang, L. Liu, F. Wang, W. Yang, G. Zheng, D. Wu, Z. Leung, M. K. 2014. A high-capacity dual-electrolyte aluminum/air electrochemical cell. *RSC Advances* 4: 30857–30863.
26. Sun, Y. Fu, C. Ren, J. Jiang, M. Guo, M. Cheng, R. Zhang, J. Sun, B. 2020. Highly conductive and reusable electrolyte based on sodium polyacrylate composite for flexible Al-air batteries. *Journal of the Electrochemical Society* 167: 080502.
27. Shui, Z. Liao, X. Lei, Y. Ni, J. Liu, Y. Dan, Y. Zhao, W. Chen, X. 2020. MnO_2 synergized with N/S codoped graphene as a flexible cathode efficient electrocatalyst for advanced honeycomb-shaped stretchable aluminum-air batteries. *Langmuir* 36: 12954–12962.
28. Xu, Y. Zhao, Y. Ren, J. Zhang, Y. Peng, H. 2016. An all-solid-state fiber-shaped aluminum-air battery with flexibility, stretchability, and high electrochemical performance. *Angewandte Chemie International Edition* 128: 8111–8114.

CHAPTER 9

Stretchable Batteries

9.1 OVERVIEW OF STRETCHABLE BATTERIES

Wearable electronics have experienced a booming development in the past decade, which can be integrated with clothes, attached on the skin, and implanted into organs to achieve a variety of applications, e.g., mobile communication, real-time health care, and human-computer interaction [1–4]. In addition to meeting the requirements of flexibility, wearable electronics should also be stretchable, so that they can be compatible with the deformations of our body during movement [5–8]. To this end, the development of suitable batteries with stretchability has become an urgent need in the field of energy storage devices.

Compared with flexible batteries, stretchable batteries must realize stable electrochemical performances not only under bending or twisting but also under tensile deformation [9–11]. This puts forward more requirements for its structure of devices and composition of materials. Recently, there are two general strategies for constructing stretchable batteries. The first one is to use intrinsically stretchable components including electrodes, electrolyte, and packaging materials. The second one is to achieve stretchability by special structure design including wavy, serpentine, kirigami-patterned, and spring-like architectures [12].

9.2 INTRINSICALLY STRETCHABLE BATTERIES

To fabricate intrinsically stretchable battery, all its components, including electrodes and electrolyte, should be stretchable. Among them, fabricating intrinsically stretchable electrodes, which consist of current collectors and

DOI: 10.1201/9781003273677-9

active materials, is the key to produce an intrinsically stretchable battery. Two approaches, namely, coating active materials on conductive elastic current collectors or embedding active materials inside porous stretchable current collectors, are commonly used to produce intrinsically stretchable electrodes [13–17]. As for the electrolyte, liquid electrolyte can be well sealed during packing or using gel electrolyte to achieve stretchability [14,17].

Carbon materials, including carbon nanotubes (CNTs), carbon black, and graphene, are the most used materials for producing stretchable current collectors because of their relatively high conductivity, large specific surface area, high chemical stability, easy manufacture, and low cost. For example, CNT and carbon black were used as conductive fillers dispersed in hexane before being homogeneously mixed with an ecoflex matrix (Figure 9.1a) [14]. After adding the curing agent, the hybrids were poured into a petri dish for drying. Finally, an elastic current collector was obtained. The current collector showed a self-assembled structure similar to Jabuticaba tree (Figure 9.1b). The carbon black nanoparticles adhered

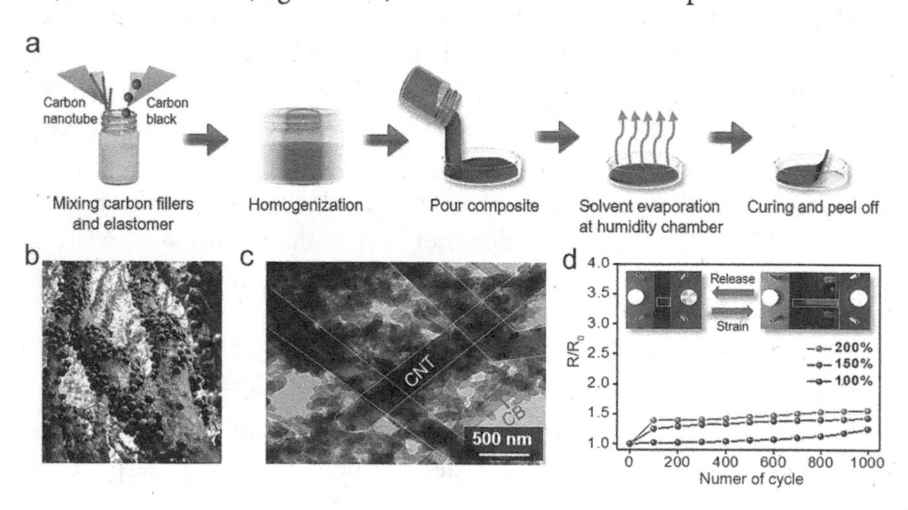

FIGURE 9.1 (a) Schematic illustration to the fabrication process of a stretchable current collector using carbon nanotubes (CNTs) and carbon black conductive fillers. (b) Photograph of a Jabuticaba tree, which has a similar shape to that of the percolating network of the hybrid carbon. (c) Transmission electron microscopy (TEM) image of the hybrid conductive polymer composite. (d) Normalized resistance of the hybrid carbon/polymer (HCP) composite under strains of 100%, 150%, and 200% that were repeated for 1,000 cycles. (Reproduced from Ref. [14] with permission of Wiley-VCH.)

to the surface of the CNT tree and occupied the space in CNT networks (Figure 9.1c). During stretching, the nonclustered carbon black particles in the polymer matrix can bridge the gaps among clustered conductive fillers. As a result, the hybrid carbon clusters can be well interconnected with each other upon stretching, ensuring good electrically conductive channels. Therefore, the elastic current collector performed stable sheet resistance, as shown in Figure 9.1d, and the sheet resistance only increased by 50% even after stretching for 1,000 cycles at stain of 200%, indicating its stable electrical conductivity performance.

The stretchable full battery was assembled by coating active materials on the hybrid conductive polymer composite current collector, and the electrodes and electrolyte were further sealed in ecoflex packing material (Figure 9.2a). The battery can be stretched by up to 100% without changing the light intensity of a red light-emitting diode (LED) (Figure 9.2b and c). Besides, the full battery exhibited a highly stable discharge capacity, and 93% of the capacity can be maintained after discharging for 500

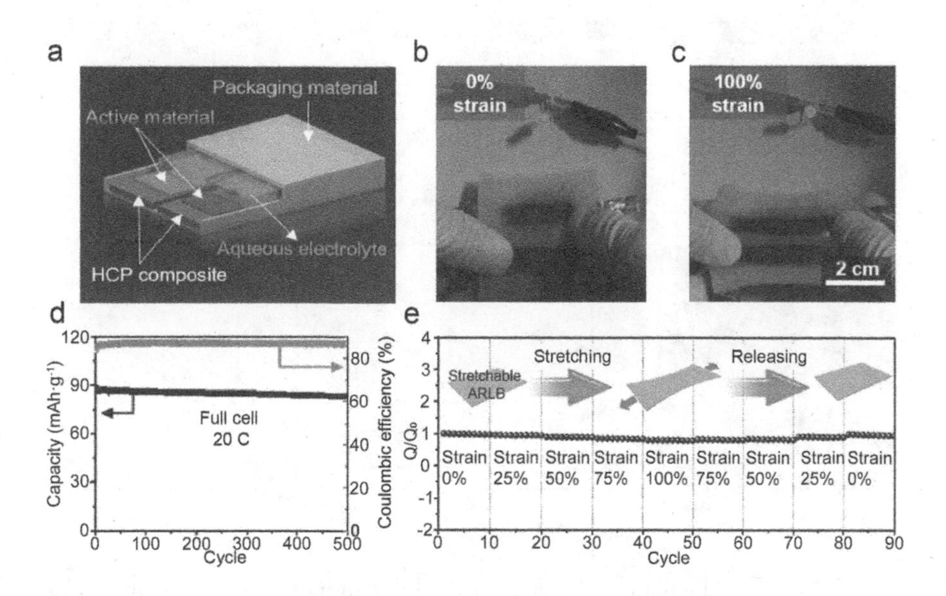

FIGURE 9.2 (a) Schematic illustration of the configuration of the stretchable aqueous full battery. (b and c) Photograph of the battery before and after being stretched to 100%, respectively. (d) Long-term cycle performance and coulombic efficiency of the full battery at a rate of 20 C over 500 cycles. (e) The relative discharge capacity of the battery under various strains. (Reproduced from Ref. [14] with permission of Wiley-VCH.)

cycles at a rate of 20 C (Figure 9.2d). The discharge performance of the battery under different strains at the rate of 20 C was also conducted and the results showed that 80% capacity can remain at the strain of 100% (Figure 9.2e). Moreover, the capacity can return to its original value after releasing the battery, revealing that this stretchable battery could accommodate the applied strains without having its electrochemical performance permanently impaired.

Although stretchable current collectors with carbon materials show stable conductivity, the sheet resistance is relatively high. Replacing/incorporating carbon materials with metals as conductive fillers can improve the conductivity of the current collector [15,16]. For example, after covered with a thin layer of Ag flakes on the surface of the current collector using carbon conductive fillers, the sheet resistance can be significantly improved (Figure 9.3a) [15]. At the unstretched state, the thick

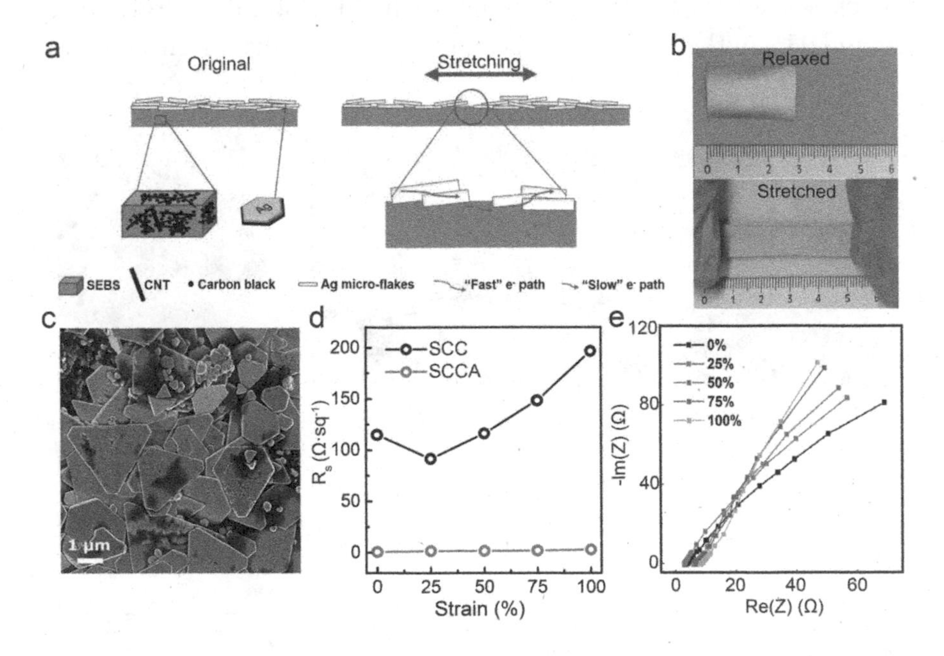

FIGURE 9.3 (a) Schematic representation of the conduction mechanism of the Ag flake-coated composite current collector at unstretched and stretched states, illustrating the contribution of the different components to the electron percolation pathways. (b) Photographs of the unstretched and stretched current collector. (c) SEM image of the current collector. (d) Sheet resistance of the current collector at various strains. (e) Nyquist plots of the stretchable battery at various strains. (Reproduced from Ref. [16] by permission of Wiley-VCH.)

Ag flakes covered on the surface of current collector, forming a continuous and fast electron percolation pathway. Upon stretching, the Ag flakes were pulled apart in stretching direction with thickness and contact area reduced, which slightly increased the resistance of the current collector. Despite local cracks that may appear on the current collector, the electrons can be transferred through the backup conductive substrate, enabling the high conductivity of the current collector. Figure 9.3b and c show the photograph and scanning electron microscopy (SEM) image of the stretchable current collector. The flakes formed a porous, continuous multilayered conductive film on the substrate. As a result, the current collector with the Ag layer owned a quite low sheet resistance of 0.53 $\Omega \cdot sq^{-1}$ at the initial state (Figure 9.3d) while the one without the Ag layer showed a much high sheet resistance of 115.0 $\Omega \cdot sq^{-1}$. After being stretched to 100%, the sheet resistance of the current collector with Ag flakes only increased to 2.71 $\Omega \cdot sq^{-1}$, which is much smaller than those using bare carbon materials. A stretchable full battery was assembled by sandwiching stretchable gel electrolyte between two active material-coated current collectors. The Nyquist plots of the stretchable battery at various strains performed similar impedances, which further proved the conductivity of the current collector (Figure 9.3e).

Coating active materials on the surface of elastic current collector to assemble the stretchable battery has the advantage of simple fabrication. However, the produced battery commonly exhibits low capacity retention after repeated stretching cycles because the active materials easily peel off from the current collector. Embedding active materials in a porous stretchable current collector can efficiently solve this problem [13,17]. For example, by using sugar cube as the pore-creating agent, a porous sponge-like polydimethylsiloxane (PDMS) scaffold with high stretchability was fabricated (Figure 9.4a) [13]. The stretchable electrodes were then prepared by filling electrode materials including active materials ($Li_4Ti_5O_{12}$ and $LiFePO_4$), carbon black, and binder in the sponges. Figure 9.4b shows the SEM image of the sugar cube, indicating that the microstructure of the sponge was the inverse matrix of the sugar, in which the pore distribution and orientation followed the sugar templates. Besides, the pore size and distribution of the PDMS sponge can be modified by choosing sugar cubes with different sizes. The stretchable electrodes could be stretched by 80% without breakage due to their robust three-dimensional interconnected porous structure. The discharge profiles in Figure 9.4c and d proved the

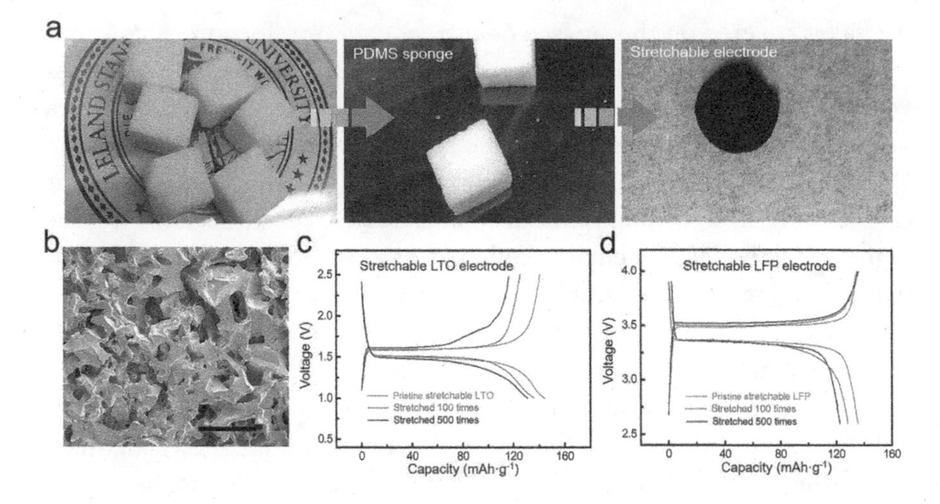

FIGURE 9.4 (a) Photographs of sugar cube, polydimethylsiloxane (PDMS) sponge, and stretchable electrode. (b) SEM image of the commercial sugar cubes. (c and d) Charge and discharge profiles of the electrode after various stretching and releasing cycles, respectively. (Reproduced from Ref. [13] with permission of Wiley-VCH.)

capacity of the electrode can be maintained over 80% and 90% after stretching for 500 cycles at the strain of 50% for the $Li_4Ti_5O_{12}$ anode and $LiFePO_4$ cathode, respectively. The stable discharge properties of the electrodes were mainly attributed to the fact that its porous structure can efficiently avoid the active materials to detach from the current collector, thus enabling high capacity retention. Although embedding electrode materials in the stretchable PDMS sponge can efficiently prevent the electrode materials from detaching from the current collector, the shortage of this method is also clear, i.e., the PDMS sponge is nonconductive. Using conductive materials to produce stretchable current collectors with porous structures is more favorable for the batteries, which need to pay more efforts in the future.

9.3 STRUCTURALLY STRETCHABLE BATTERIES

Although it is possible to fabricate stretchable batteries by enabling its components to be elastic, the strain of the intrinsically stretchable battery is commonly low. To this end, four kinds of stretchable engineering strategies, namely wavy, serpentine, kirigami pattern, and spring like, have been adopted to fabricate stretchable battery with larger stretchability.

9.3.1 Wavy Stretchable Batteries

The wavy structure is a commonly used strategy to fabricate thin-film stretchable batteries [18–21]. Generally, the wavy shape battery is made of two stretchable wavy electrodes and a gel electrolyte. The wavy shape electrodes are produced by coating electrode materials on the pre-strained elastomers, followed by releasing them to their initial states. The whole battery then can be assembled by sandwiching the gel electrolyte between the two electrodes. The tensile strains of the wavy shape batteries are achieved mainly through the change in the amplitude and wavelength of the wavy shape, and one representative analytic model based on the theory of finite deformation has pointed out the stretchability of the electrodes increases with increasing pre-strains [22,23].

Figure 9.5a shows the typical structure of a stretchable battery with wavy electrodes [18]. The stretchable gel electrolyte was sandwiched between two electrodes and the entire battery was encapsulated in PDMS. The electrodes were prepared by sandwiching lithium manganese oxide and lithium titanium oxide between conductive and flexible CNT sheets initially. Then the electrode was transferred on a pre-strained elastic substrate. After releasing the elastic polymer to its initial state, a stretchable electrode with wavy shape has been prepared (Figure 9.5b). The produced wavy geometry of the electrode guaranteed the battery stretchable with the maintained sandwich structure, and the active materials were stably anchored in the CNT sheets without peeling off from the electrode even after stretching for hundreds of cycles. As a result, the battery had high flexibility and stable electrochemical performance. The output energy of the assembled battery remained almost unchanged after stretching to 400%, and 97% of it could be maintained even after the battery had been stretched for 200 cycles at the strain of 400% (Figure 9.5c and d). The light intensity of the lighted LED remained stable after the battery was stretched to 400% strain, further indicating the stable output of the battery (Figure 9.5e and f). Based on the same strategy, transferring electrode materials on the biaxially pre-strained elastomer can produce the biaxially stretchable battery. For example, by transferring the above used sandwiched CNT/active material on a biaxially pre-strained PDMS, a battery with biaxial stretchability had been achieved [19]. It can be stretched to 150% in different axes without obvious degradation in performance.

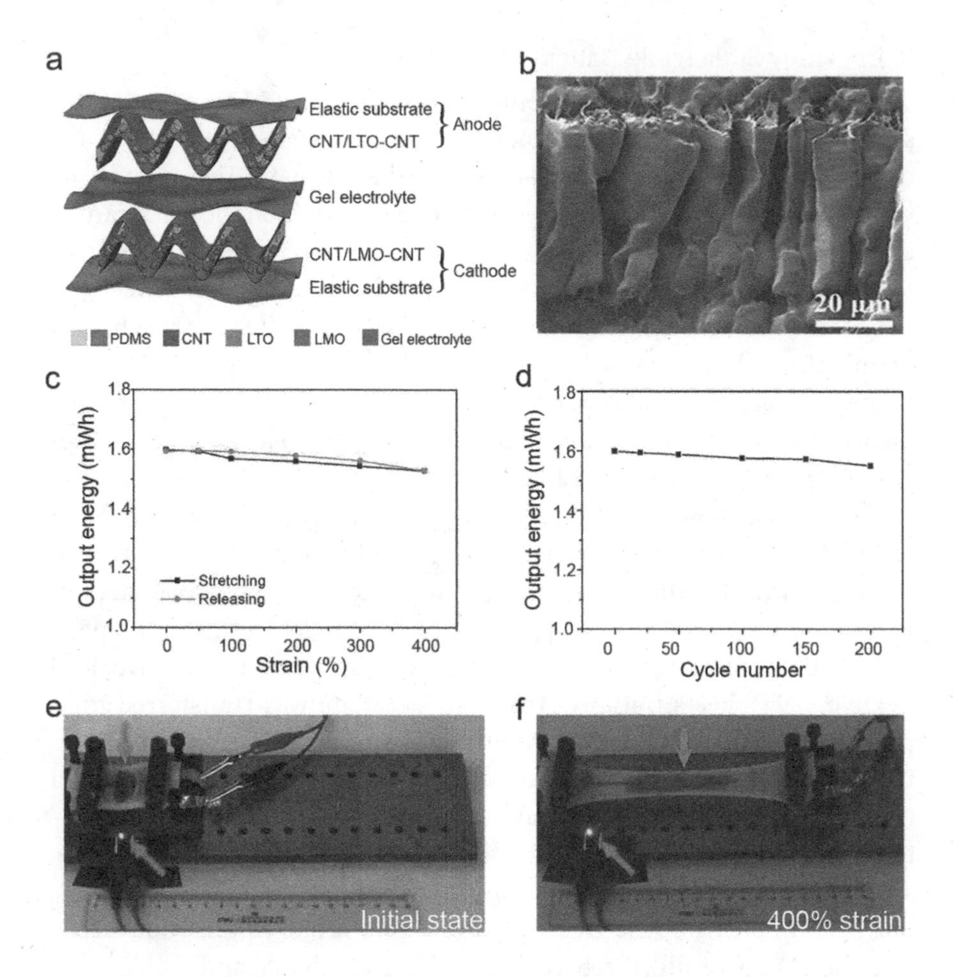

FIGURE 9.5 (a) Schematic illustration of a stretchable battery with wavy shape electrode. (b) SEM image of the wavy electrode. (c and d) Dependence of the output energy on stretching strain and stretching cycle number, respectively. (e and f) Photographs of the stretchable battery at the initial and 400% strain state, respectively. (Reproduced from Ref. [18] with permission of Wiley-VCH.)

9.3.2 Serpentine Stretchable Batteries

Different from the wavy stretchable battery, serpentine stretchable battery commonly consists of rigid energy storage units and stretchable serpentine conductive wires, both of which are mounted or embedded in an elastomer [24–26]. The stretchability of the serpentine stretchable battery is mainly achieved through the in-plane deformation of the serpentine conductive wires rather than the functional energy storage units, and thus the geometry of the serpentine conductive wires is significantly important.

Four independent parameters, including arc radius, width, arm length, and arc opening angle, are widely used to describe the geometry of the serpentine conductive wires and different combinations of these parameters can induce different geometries for various applications [7,27,28]. Although the deformation behaviors of the serpentine conductive wires are quite different, the stretchability of the entire device can be briefly evaluated by the formula of

$$\varepsilon_{\text{device}} = \left(1 - \frac{\sqrt{S_a}}{\sqrt{S_t}}\right)\varepsilon_{\text{serpentine wires}}$$

where $\varepsilon_{\text{device}}$ and $\varepsilon_{\text{serpentine wires}}$ are the stretchability of the entire battery and serpentine wires; S_a and S_t are the areas of functional energy storage units and the total area of the device, respectively [29,30]. Obviously, the stretchability of the final battery is inversely proportional to the area of the functional energy storage unit. For this reason, the serpentine conductive wires should be specifically designed to balance energy storage function and stretchability.

The representative work to fabricate stretchable batteries using serpentine conductive wires was shown in Figure 9.6a and b [24]. The resulting stretchable battery system showed 100 energy storage units. The entire battery was laminated in an elastomer to obtain support and protection. Figure 9.6c exhibits the carefully designed two-level serpentine conductive wires, and each of them was connected with two energy storage units (Figure 9.6d). The fabricated battery can be stretched by 300%, and the discharge and charge profiles are similar before and after stretching (Figure 9.6e). Moreover, the battery showed little capacity degradation after cycling for 20 cycles (Figure 9.6f). Figure 9.6g presents the photograph of the fabricated battery, and the red LED still can be lighted after being stretched to 300%, proving the stretchability of the battery (Figure 9.6h).

9.3.3 Kirigami-Patterned Structure

Kirigami-pattern structure, inspired from ancient paper arts and combines folding, cutting, and other manufacturing technologies, has been recently used to achieve the out-of-plane deformations of two-dimensional building blocks [31]. Similar to the stretchable serpentine structure, the kirigami pattern commonly consists of many cuts with periodic parallel distributions. Differently, the stretchability of the kirigami pattern is achieved through the deformation of long "bridges," rather than boned

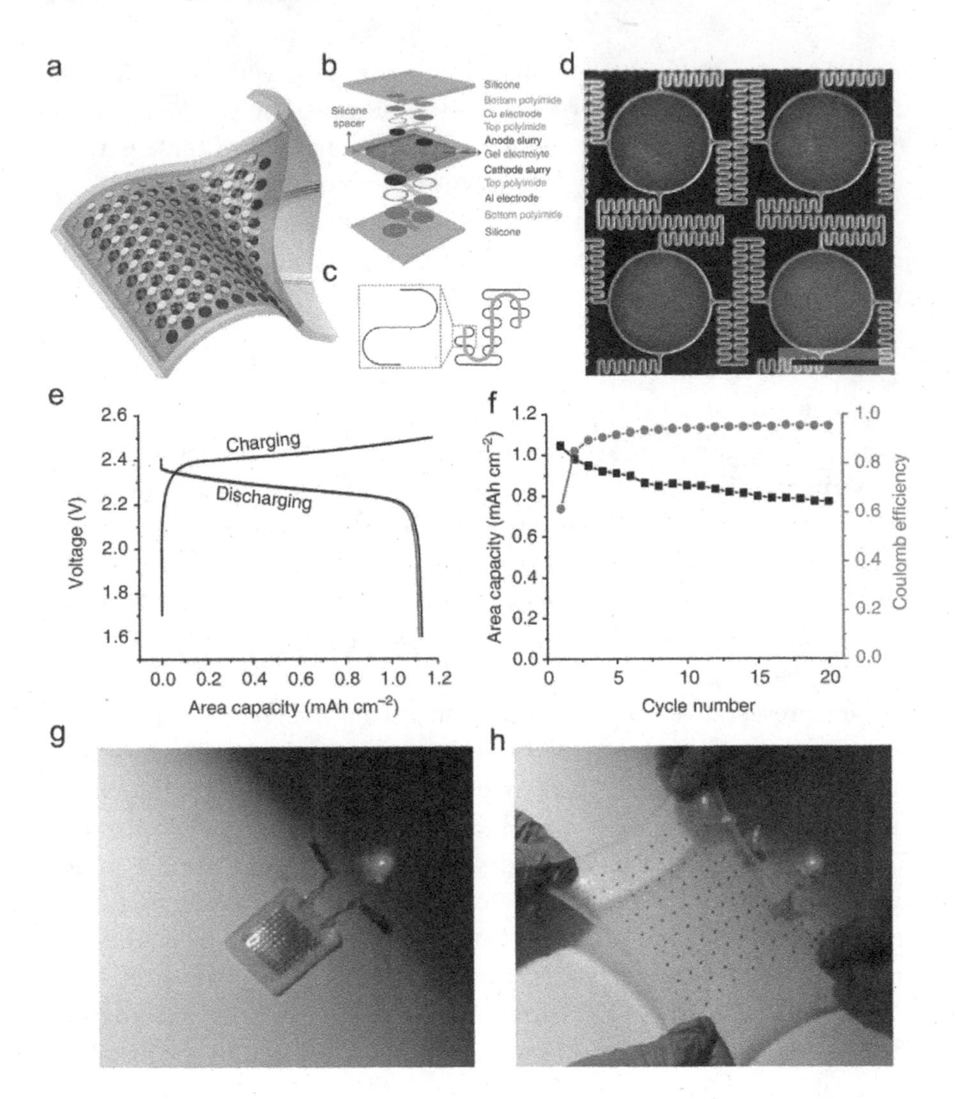

FIGURE 9.6 (a) Schematic illustration of a battery at stretching and bending states. (b) Exploded view layout of the various layers in the battery. (c) Illustration of "self-similar" serpentine geometries used for the interconnects (black: first level serpentine; yellow: second level serpentine). (d) Optical images of the Al electrode pads and self-similar interconnects on a Si wafer. (e) Galvanostatic charging and discharging curves of the battery electrodes without (black) and with 300% uniaxial strain (red). (f) Capacity retention (black) and coulombic efficiency (red) over 20 cycles with a cutoff voltage of 2.5–1.6 V. (g and h) Operation of a battery connected to a red LED before and after stretching, respectively. (Reproduced from Ref. [24] with permission of Nature Publishing Group.)

stretchable serpentine wires. Therefore, there is no need to bond the "bridges" on the elastic substrate and the deformations of kirigami pattern thus are not limited in plane.

One of the representative stretchable kirigami patterns with periodic parallel distributions before and after stretching is presented in Figure 9.7a and b [32]. The stretchability of the kirigami-pattern structure can be evaluated by a beam model with a slit design [32,33]. Ignoring the out-of-plane bending of each beam with a rotation, the stretchability of the film is proportional to the total load of the film, which can be expressed as

$$d_{\text{total}} = \frac{N_c P\left(L_{\text{slit}} - L_{\text{gap}}\right)^3}{8N_r Ewt^3}$$

where N_r and N_c are the row and column numbers of the battery units, respectively. E, w, and t are Young's modulus, beam width, and film thickness, respectively; L_{slit} and L_{gap} are slit length and gap between the slits, respectively (see Figure 9.7c). According to this formula, the bending property of each unit contributes to the total stretchability of the film and it mainly depends on the material (E) and geometry (e.g., w and t).

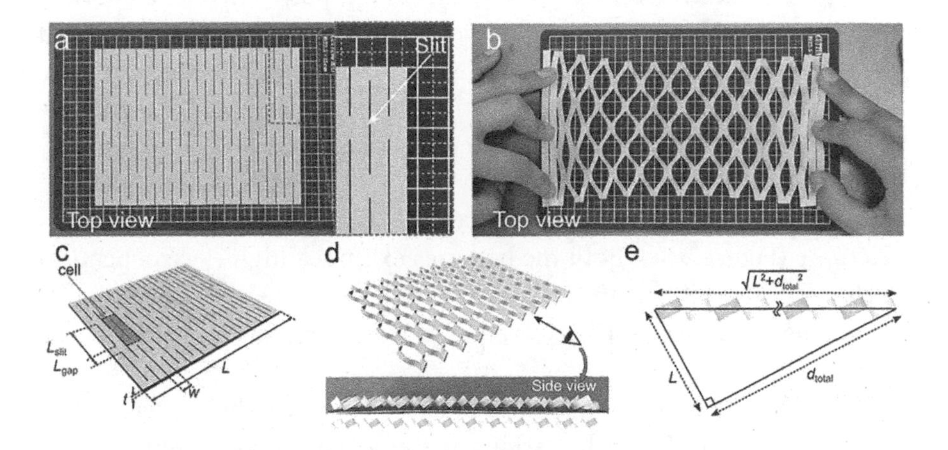

FIGURE 9.7 (a and b) Photographs of the kirigami-pattern structure before and after stretching, respectively. (c) Schematic illustration of a stretchable unit used for analyzing its stretchability. (d) Schematic illustration and photograph of the out-of-plane deformation based on stretchable kirigami pattern. (e) Accurate displacement of the stretched film by considering the Pythagorean theorem. (Reproduced from Ref. [32] with permission of Wiley-VCH.)

Also, the stretchability of the film is related to the column/row numbers (N_c/N_r ratio). When taking the rotation of each beam into account, the stretchability of the film d_{actual} can be given by the Pythagorean theorem (Figure 9.7d and e) as

$$d_{\text{actual}} = \sqrt{L^2 + d_{\text{total}}^2} - L$$

where L is the film length before stretching. This formula is more accurate when a large deformation that leads to the beam unit rotation happens. Besides, the finite element analysis proved even at a large deformation, the strains can be uniformly dissipated in most regions of the whole pattern. Thus, although the connection point may show relatively large stain concentrations, the whole battery can maintain stability after repeat stretching.

To prepare the stretchable battery with kirigami pattern, the first thing is to produce a stretchable kirigami-patterned electrode. Figure 9.8a shows the schematic illustration of two kinds of kirigami patterns for constructing stretchable electrodes [34]. The electrodes were prepared using an evolutionary template method and universal electrode inks, which consist of active materials ($LiFePO_4$ and $Li_4Ti_5O_{12}$), poly(vinylidene fluoride) binder, and CNT conductive additive. After fully drying, the produced electrode was flexible and stretchable (Figure 9.8b). Compared with the conventional electrodes that presented a bilayer structure (active materials coated on the current collector), the mixed electrode ink can effectively avoid interface debonding and alleviate structure damages under the stretching condition. The full punch cell was assembled with the produced electrode (Figure 9.8c). Both the batteries exhibited high electrochemical performance, especially the battery with the edge-cut electrode (design 1) maintained over 80% capacity even after stretching for 500 cycles at the strain of 100% (Figure 9.8d).

Figure 9.9 shows the photographs of several stretchable batteries with kirigami patterns [34–37]. Various patterns enabled different battery strain ranges and orientations. For example, the battery in Figure 9.9a can be stretched by 1,600% by folding and unfolding the kirigami pattern and the battery in Figure 9.9b can bear the perpendicular strain. Thus, similar to the stretchable serpentine battery, the kirigami pattern should be well designed and considered to achieve specific applications.

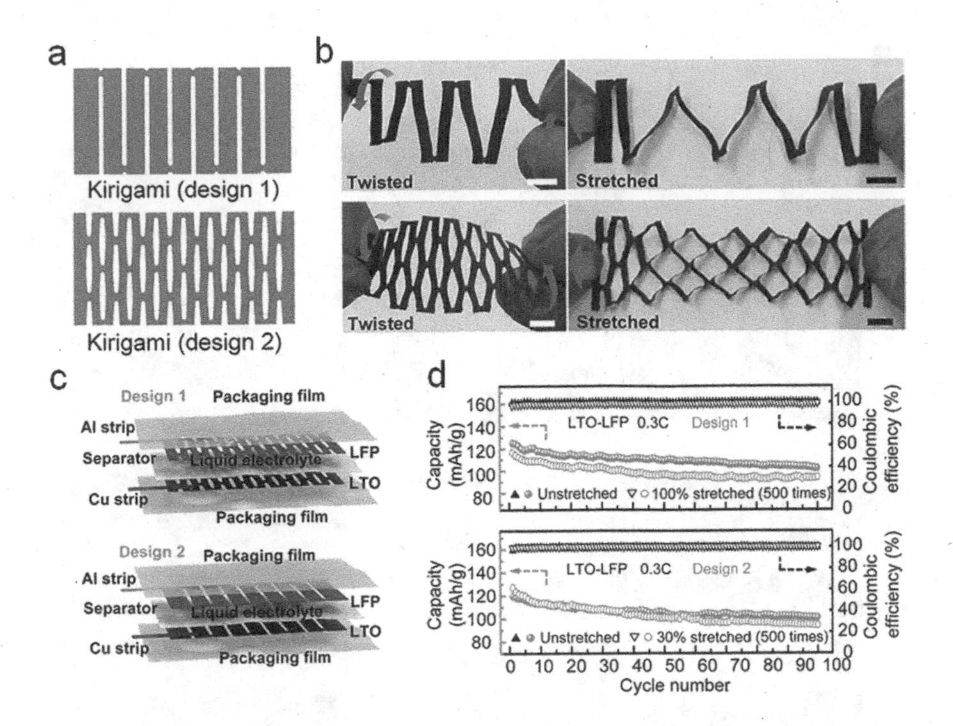

FIGURE 9.8 (a) Schematic illustration of two stretchable kirigami patterns. (b) Photographs of two designed kirigami $LiFePO_4$ cathodes at twisted and stretched states. (c) Illustration of the full punch cell with kirigami electrodes. (d) Cycling performance and coulombic efficiency of the pouch cell with kirigami electrodes. (Reproduced from Ref. [34] with permission of American Chemical Society.)

9.3.4 Spring-Like Structure

Fiber devices attract increasing interests in the recent decade due to their lightweight, flexible, weaveable, and breathable features. Moreover, they can be easily integrated with a cloth to achieve wearable electronics, which is the developing trend of modern electronics. Stretchable fiber Li-ion batteries also have been reported to satisfy the demand for powering wearable electronics.

To fabricate a fiber stretchable Li-ion battery, spring-like fiber electrodes are the keys. Commonly used active materials, such as $LiMn_2O_4$ and $Li_4Ti_5O_{12}$, are particles, and they thus have to combine with a scaffold to produce a fiber electrode. The dry-drawn CNT sheets, which have the characteristics of being flexible, conductive, lightweight, and porous, are widely used as the scaffold to produce the spring-like electrodes.

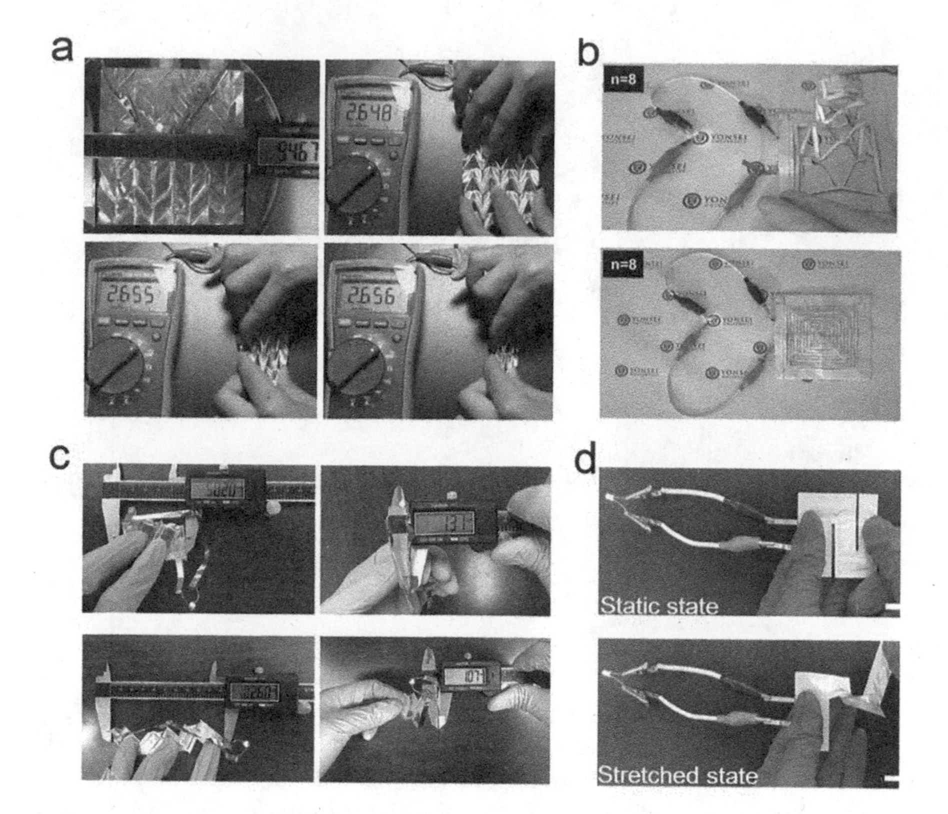

FIGURE 9.9 Photographs of stretchable batteries with different kirigami patterns. (a) The battery can be stretched by 1,600%, reproduced from Ref. [36] with permission of Nature Publishing Group. (b) A battery can be stretched in direction perpendicular to the plane, reproduced from Ref. [37] with permission of Wiley-VCH. (c) Photographs of stretchable batteries with cut-N-twist and cut-N-shear patterns at their compact and stretched states, respectively [35] (d) Photograph of a battery with edge-cut kirigami pattern at its static and stretched states. (Reproduced from Ref. [34] with permission of American Chemical Society.)

Figure 9.10a schematically presents the fabrication process of a stretchable fiber Li-ion battery [38]. Active materials were firstly coated on the dry-drawn CNT sheets, and then the sheets were twisted into fiber shape electrodes. The electrodes were further paralleled and wound on an elastic substrate to achieve a stretchable spring-like structure. Lastly, a thin layer of gel electrolyte and a thin layer of PDMS were sequentially coated to seal the battery. The active materials were uniformly wrapped into the CNT fiber electrode with a porous structure, which is beneficial for increasing

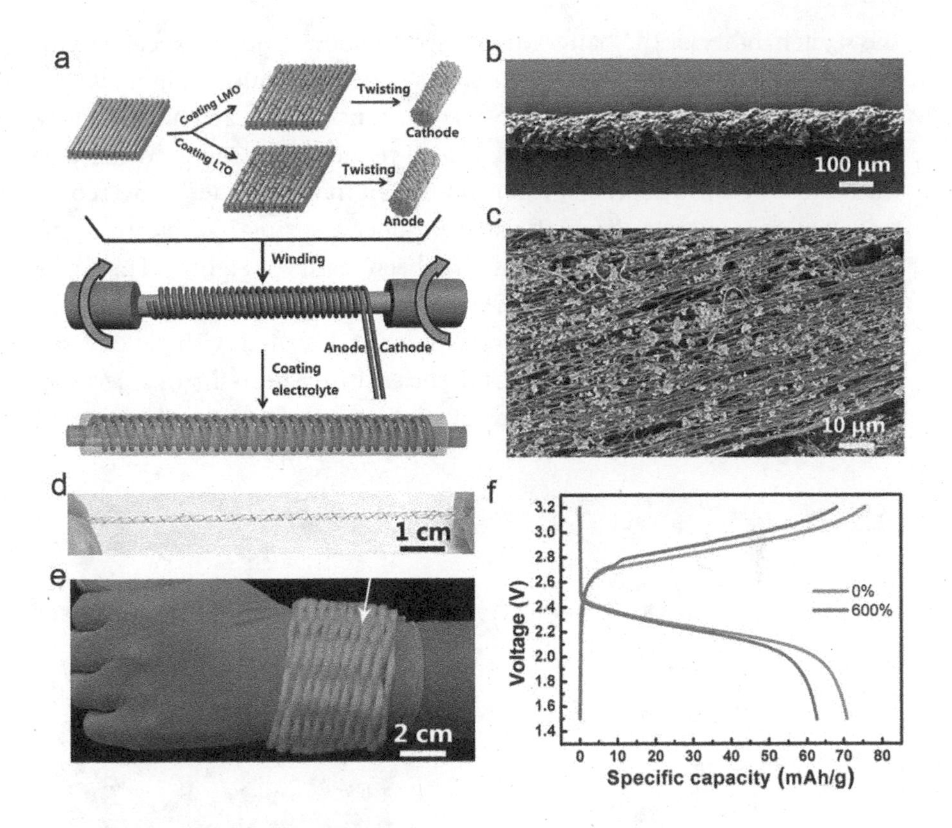

FIGURE 9.10 (a) Schematic illustration of the fabrication of the stretchable Li-ion battery based on the twisted carbon nanotube (CNT) sheets and $LiMn_2O_4$ composite fiber as a positive electrode and twisted CNT sheets and $Li_4Ti_5O_{12}$ composite fiber as a negative electrode. (b and c) Low and high magnification SEM images of the fiber electrode, respectively. (d) Photograph of the stretchable battery after being stretched to 200%. (e) The fiber battery can be woven into a bracelet textile for powering wearable electronic devices. (f) Charge and discharge profiles of the fiber battery before and after stretching to 600% at a current density of 0.1 $mA{\cdot}cm^{-1}$. (Reproduced from Ref. [38] with permission of the Royal Society of Chemistry.)

the contact area between the electrode and the electrolyte, thus favoring the charge transfer (Figure 9.10b and c). Due to the spring-like structure and unique one-dimensional construction, the battery was stretchable (200%) and could be woven into a bracelet textile (Figure 9.10d and e). Besides, the stretchability of the fiber battery can be easily adjusted by changing the pitch distance between the two electrodes and the diameter of the elastic substrate. For example, by changing the above parameters, the stretchability of the battery increased to 600%, and the specificity of the battery was maintained by over 88% at a strain of 600% (Figure 9.10f).

The merit of using an elastic substrate to construct a stretchable battery is its high stretchability. However, it also reduced the energy density of the battery because its weight is far more than the electrode. To overcome this challenge, using an overtwisted-produced spring-like electrode to assemble the stretchable battery also has been realized. Figure 9.11a shows the structure of the battery with overtwisted spring-like electrode [39]. To assemble a full battery, two electrodes were first coated with gel electrolyte, and they were then twisted and encased in a heat-shrinkable tube. The coiled loops were gradually elongated during the stretch and returned

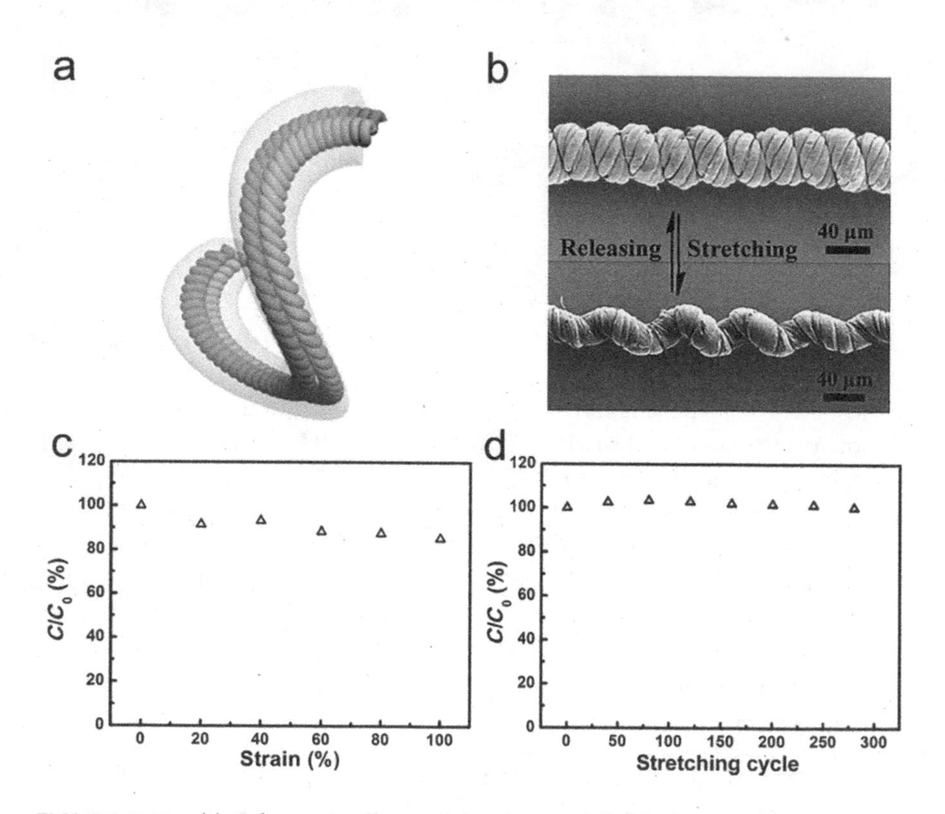

FIGURE 9.11 (a) Schematic illustration of a stretchable battery based on the over-twisted spring-like electrodes. (b) SEM images of the over-twisted carbon nanotube (CNT) spring at its initial and stretched states. (c) The capacity retention of the fiber battery at various strains. (d) The capacity retention of the fiber battery at the strain of 50% after various stretching cycles. (Reproduced from Ref. [39] with permission of Wiley-VCH.)

to their original coiled structure after releasing, endowing the battery stretchable (Figure 9.11b). This battery can be stretched up by 100% with 85% capacity remaining (Figure 9.11c). Besides, the capacity remained almost unchanged after repeated stretching for 300 cycles at a strain of 50%, indicating its stable electrochemical performance (Figure 9.11d). Moreover, due to the small size and light weight of the spring-like electrode, the linear specific capacity of the battery was enhanced by 600% compared to the stretchable fiber lithium-ion battery on an elastic polymer substrate [38,40].

9.4 PERSPECTIVE

During the past decades, enabling the batteries to be stretchable has attracted interest, and great efforts have thus been devoted to this field. So far, a lot of progress has been made. The stretchable batteries have been assembled from the aspects of using intrinsically stretchable materials and designing stretchable structures. The maximum stretchability of the battery has reached 1,600% to date, which is a big step for stretchable batteries. Moreover, the unique one-dimensional stretchable fiber batteries also have emerged, which is believed to have a broad application for wearable electronics. Despite these progresses, many challenges still exist before achieving the real applications.

1. Constructing stretchable battery with intrinsically stretchable materials has the advantage of easy fabrication and low cost. However, the strain ranges of the battery are commonly low, which need to be further improved to satisfy the demands of real applications. Presently, the batteries with intrinsically stretchable materials are mainly realized by coating active materials on the surface of stretchable current collectors or embedding active materials inside porous stretchable current collectors. Therefore, developing highly stretchable current collector with high conductivity is the key to produce a battery with both large stretchability and high electrochemical performance.
2. Compared to the intrinsically stretchable battery, the batteries with stretchable structures have the merits of larger stretchability and relatively stable electrochemical performances. Nevertheless, the energy density and power density of the battery with stretchable structures are commonly low due to the used elastic substrates

which are commonly several orders of magnitude heavier than the electrode materials. Using lightweight elastic substrates or even the free use of elastic substrate to fabricate the battery can improve its electrochemical performances to a certain degree. Besides, the presently used fabrication technologies are relatively complex and costly. Developing novel fabrication methods and simple fabrication processes is helpful for accelerating the real applications of the battery.

3. The electrochemical and mechanical stability of the stretchable battery needs to be further enhanced. Currently, the reported batteries only can bear several hundreds of stretching cycles. Moreover, the specific capacity and energy density of the battery showed noticeable degradation after repeated stretching/releasing cycles. The degradation mainly comes from the detachment of active materials from the current collectors, increased resistance of current collectors, and delamination of electrodes and electrolyte. Therefore, developing novel methods or structures that can tightly bond active materials on current collectors with strong adhesion between the electrodes and the electrolyte needs to be considered. Besides, fabricating a stretchable current collector that can maintain high conductivity even at the stretched state by developing new materials and assemble technologies is also necessary.
4. Last but not least, the packing materials also need to be improved. Stretchable polymers, including PDMS, ecoflex, and polyurethane, are widely used as elastic substrates and packing materials for stretchable batteries at present. However, these materials commonly showed high water and oxygen transmission rates, which led to battery failure. Thus, new packing materials with low water and oxygen transmission rates as well as high mechanical stretchability are needed.

Overall, the stretchable battery has demonstrated huge application aspects in the field of powering wearable electronic devices. However, the present progress cannot satisfy the demands of real applications. To achieve this final goal, efforts from the researchers in the field of engineering, materials sciences, electronics, and chemistry should be well joined.

REFERENCES

1. Sempionatto, J. R. Lin, M. Yin, L. De la Paz, E. Pei, K. Sonsa-Ard, T. de Loyola Silva, A. N. Khorshed, A. A. Zhang, F. Tostado, N. Xu, S. Wang, J. 2021. An epidermal patch for the simultaneous monitoring of haemodynamic and metabolic biomarkers. *Nature Biomedical Engineering* 5: 737–748.
2. Fu, X. Wang, L. Zhao, L. Yuan, Z. Zhang, Y. Wang, D. Wang, D. Li, J. Li, D. Shulga, V. Shen, G. Han, W. 2021. Controlled assembly of MXene nanoshexets as an electrode and active layer for high-performance electronic skin. *Advanced Functional Materials* 31: 2010533.
3. Wang, L. Xie, S. Wang, Z. Liu, F. Yang, Y. Tang, C. Wu, X. Liu, P. Li, Y. Saiyin, H. Zheng, S. Sun, X. Xu, F. Yu, H. Peng, H. 2020. Functionalized helical fibre bundles of carbon nanotubes as electrochemical sensors for long-term in vivo monitoring of multiple disease biomarkers. *Nature Biomedical Engineering* 4: 159–171.
4. Liu, Y. Li, J. Song, S. Kang, J. Tsao, Y. Chen, S. Mottini, V. McConnell, K. Xu, W. Zheng, Y. Tok, J. B. George, P. M. Bao, Z. 2020. Morphing electronics enable neuromodulation in growing tissue. *Nature Biotechnology* 38: 1031–1036.
5. Dai, Y. Hu, H. Wang, M. Xu, J. Wang, S. 2021. Stretchable transistors and functional circuits for human-integrated electronics. *Nature Electronics* 4: 17–29.
6. Niu, S. Matsuhisa, N. Beker, L. Li, J. Wang, S. Wang, J. Jiang, Y. Yan, X. Yun, Y. Burnetts, W. Poon, A. S. Y. Tok, J. B. H. Chen, X. Bao, Z. 2019. A wireless body area sensor network based on stretchable passive tags. *Nature Electronics* 2: 361–368.
7. Xue, Z. Song, H. Rogers, J. A. Zhang, Y. Huang, Y. 2020. Mechanically-guided structural designs in stretchable inorganic electronics. *Advanced Materials* 32: 1902254.
8. Tan, Y. Godaba, H. Chen, G. Tan, S. T. M. Wan, G. Li, G. Lee, P. M. Cai, Y. Li, S. Shepherd, R. F. Ho, J. S. Tee, B. C. K. 2020. A transparent, self-healing and high-kappa dielectric for low-field-emission stretchable optoelectronics. *Nature Materials* 19: 182–188.
9. Mackanic, D. G. Kao, M. Bao, Z. 2020. Enabling deformable and stretchable batteries. *Advanced Energy Materials* 10: 2001424.
10. Mackanic, D. G. Chang, T. Huang, Z. Cui, Y. Bao, Z. 2020. Stretchable electrochemical energy storage devices. *Chemical Society Reviews* 49: 4466–4495.
11. Choi, Y. S. Hsueh, Y. Y. Koo, J. Yang, Q. Avila, R. Hu, B. Xie, Z. Lee, G. Ning, Z. Liu, C. Xu, Y. Lee, Y. J. Zhao, W. Fang, J. Deng, Y. Lee, S. M. Vazquez-Guardado, A. Stepien, I. Yan, Y. Song, J. Haney, C. Oh, Y. S. Liu, W. Yoon, H. J. Banks, A. MacEwan, M. R. Ameer, G. A. Ray, W. Z. Huang, Y. Xie, T. Franz, C. K. Li, S. Rogers, J. A. 2020. Stretchable, dynamic covalent polymers for soft, long-lived bioresorbable electronic stimulators designed to facilitate neuromuscular regeneration. *Nature Communicatons* 11: 5990.

12. Li, L. Wang, L. Ye, T. Peng, H. Zhang, Y. 2021. Stretchable energy storage devices based on carbon materials. *Small* 17: 2005015.
13. Liu, W. Chen, Z. Zhou, G. Sun, Y. Lee, H. R. Liu, C. Yao, H. Bao, Z. Cui, Y. 2016. 3D porous sponge-inspired electrode for stretchable lithium-ion batteries. *Advanced Materials* 28: 3578–3583.
14. Song, W. J. Park, J. Kim, D. H. Bae, S. Kwak, M. J. Shin, M. Kim, S. Choi, S. Jang, J. H. Shin, T. J. Kim, S. Y. Seo, K. Park, S. 2018. Jabuticaba-inspired hybrid carbon filler/polymer electrode for use in highly stretchable aqueous Li-ion batteries. *Advanced Energy Materials* 8: 1702478.
15. Gu, M. Song, W. J. Hong, J. Kim, S. Y. Shin, T. J. Kotov, N. A. Park, S. Kim, B. S. 2019. Stretchable batteries with gradient multilayer conductors. *Science Advances* 5: eaaw1879.
16. Chen, X. Huang, H. Pan, L. Liu, T. Niederberger, M. 2019. Fully integrated design of a stretchable solid-state lithium-ion full battery. *Advanced Materials* 31: 1904648.
17. Kang, S. Hong, S. Y. Kim, N. Oh, J. Park, M. Chung, K. Y. Lee, S. S. Lee, J. Son, J. G. 2020. Stretchable lithium-ion battery based on re-entrant micro-honeycomb electrodes and cross-linked gel electrolyte. *ACS Nano* 14: 3660–3668.
18. Weng, W. Sun, Q. Zhang, Y. He, S. Wu, Q. Deng, J. Fang, X. Guan, G. Ren, J. Peng, H. 2015. A gum-like lithium-ion battery based on a novel arched structure. *Advanced Materials* 27: 1363–1369.
19. Yu, Y. Luo, Y. Wu, H. Jiang, K. Li, Q. Fan, S. Li, J. Wang, J. 2018. Ultrastretchable carbon nanotube composite electrodes for flexible lithium-ion batteries. *Nanoscale* 10: 19972–19978.
20. Liu, W. Chen, J. Chen, Z. Liu, K. Zhou, G. Sun, Y. Song, M.-S. Bao, Z. Cui, Y. 2017. Stretchable lithium-ion batteries enabled by device-scaled wavy structure and elastic-sticky separator. *Advanced Energy Materials* 7: 1701076.
21. Wang, C. Zheng, W. Yue, Z. Too, C. O. Wallace, G. G. 2011. Buckled, stretchable polypyrrole electrodes for battery applications. *Advanced Materials* 23: 3580–3584.
22. Song, J. Jiang, H. Liu, Z. J. Khang, D. Y. Huang, Y. Rogers, J. A. Lu, C. Koh, C. G. 2008. Buckling of a stiff thin film on a compliant substrate in large deformation. *International Journal of Solids and Structures* 45: 3107–3121.
23. Jiang, H. Khang, D.-Y. Song, J. Sun, Y. Huang, Y. Rogers, J. A. 2007. Finite deformation mechanics in buckled thin films on compliant supports. *Proceedings of the National Academy of Sciences* 104: 15607–15612.
24. Xu, S. Zhang, Y. Cho, J. Lee, J. Huang, X. Jia, L. Fan, J. A. Su, Y. Su, J. Zhang, H. Cheng, H. Lu, B. Yu, C. Chuang, C. Kim, T. I. Song, T. Shigeta, K. Kang, S. Dagdeviren, C. Petrov, I. Braun, P. V. Huang, Y. Paik, U. Rogers, J. A. 2013. Stretchable batteries with self-similar serpentine interconnects and integrated wireless recharging systems. *Nature Communications* 4: 1543.
25. Yin, L. Seo, J. K. Kurniawan, J. Kumar, R. Lv, J. Xie, L. Liu, X. Xu, S. Meng, Y. S. Wang, J. 2018. Highly stable battery pack via insulated, reinforced, buckling-enabled interconnect array. *Small* 14: 1800938.

26. Qu, S. Song, Z. Liu, J. Li, Y. Kou, Y. Ma, C. Han, X. Deng, Y. Zhao, N. Hu, W. Zhong, C. 2017. Electrochemical approach to prepare integrated air electrodes for highly stretchable zinc-air battery array with tunable output voltage and current for wearable electronics. *Nano Energy* 39: 101–110.
27. Fan, Z. Zhang, Y. Ma, Q. Zhang, F. Fu, H. Hwang, K.-C. Huang, Y. 2016. A finite deformation model of planar serpentine interconnects for stretchable electronics. *International Journal of Solids and Structures* 91: 46–54.
28. Xu, L. Gutbrod, S. R. Bonifas, A. P. Su, Y. Sulkin, M. S. Lu, N. Chung, H.-J. Jang, K.-I. Liu, Z. Ying, M. Lu, C. Webb, R. C. Kim, J.-S. Laughner, J. I. Cheng, H. Liu, Y. Ameen, A. Jeong, J.-W. Kim, G.-T. Huang, Y. Efimov, I. R. Rogers, J. A. 2014. 3D multifunctional integumentary membranes for spatiotemporal cardiac measurements and stimulation across the entire epicardium. *Nature Communications* 5: 3329.
29. Su, Y. Ping, X. Yu, K. J. Lee, J. W. Fan, J. A. Wang, B. Li, M. Li, R. Harburg, D. V. Huang, Y. Yu, C. Mao, S. Shim, J. Yang, Q. Lee, P. Y. Armonas, A. Choi, K. J. Yang, Y. Paik, U. Chang, T. Dawidczyk, T. J. Huang, Y. Wang, S. Rogers, J. A. 2017. In-plane deformation mechanics for highly stretchable electronics. *Advanced Materials* 29: 1604989.
30. Zhang, Y. Fu, H. Xu, S. Fan, J. A. Hwang, K. C. Jiang, J. Rogers, J. A. Huang, Y. 2014. A hierarchical computational model for stretchable interconnects with fractal-inspired designs. *Journal of the Mechanics and Physics of Solids* 72: 115–130.
31. Xu, L. Shyu, T. C. Kotov, N. A. 2017. Origami and kirigami nanocomposites. *ACS Nano* 11: 7587–7599.
32. Morikawa, Y. Yamagiwa, S. Sawahata, H. Numano, R. Koida, K. Ishida, M. Kawano, T. 2018. Ultrastretchable kirigami bioprobes. *Advanced Healthcare Materials* 7: 1701100.
33. Shyu, T. C. Damasceno, P. F. Dodd, P. M. Lamoureux, A. Xu, L. Shlian, M. Shtein, M. Glotzer, S. C. Kotov, N. A. 2015. A kirigami approach to engineering elasticity in nanocomposites through patterned defects. *Nature Materials* 14: 785–789.
34. Bao, Y. Hong, G. Chen, Y. Chen, J. Chen, H. Song, W. L. Fang, D. 2020. Customized kirigami electrodes for flexible and deformable lithium-ion batteries. *ACS Applied Materials & Interfaces* 12: 780–788.
35. Song, Z. Wang, X. Lv, C. An, Y. Liang, M. Ma, T. He, D. Zheng, Y. J. Huang, S. Q. Yu, H. Jiang, H. 2015. Kirigami-based stretchable lithium-ion batteries. *Scientific Reports* 5: 10988.
36. Song, Z. Ma, T. Tang, R. Cheng, Q. Wang, X. Krishnaraju, D. Panat, R. Chan, C. K. Yu, H. Jiang, H. 2014. Origami lithium-ion batteries. *Nature Communications* 5: 3140.
37. Choi, S. Lee, D. Kim, G. Lee, Y. Y. Kim, B. Moon, J. Shim, W. 2017. Shape-reconfigurable aluminum-air batteries. *Advanced Functional Materials* 27: 1702244.

38. Zhang, Y. Bai, W. Ren, J. Weng, W. Lin, H. Zhang, Z. Peng, H. 2014. Superstretchy lithium-ion battery based on carbon nanotube fiber. *Journal of Materials Chemistry A* 2: 11054–11059.
39. Zhang, Y. Bai, W. Cheng, X. Ren, J. Weng, W. Chen, P. Fang, X. Zhang, Z. Peng, H. 2014. Flexible and stretchable lithium-ion batteries and supercapacitors based on electrically conducting carbon nanotube fiber springs. *Angewandte Chemie International Edition* 53: 14564–14568.
40. Ren, J. Zhang, Y. Bai, W. Chen, X. Zhang, Z. Fang, X. Weng, W. Wang, Y. Peng, H. 2014. Elastic and wearable wire-shaped lithium-ion battery with high electrochemical performance. *Angewandte Chemie International Edition* 53: 7864–7869.

CHAPTER 10

Integration of Flexible Batteries

10.1 OVERVIEW OF INTEGRATED DEVICES

Flexible and wearable electronic devices represent a mainstream direction in modern electronics, which are currently changing the face of mobile communication, health care, human-computer interaction, etc. [1–3]. To achieve systematic functional applications while easing the installation, different electronic components such as batteries and sensors need to be integrated [4]. Although the integration technology of the traditional electronic device is mature, it is not suitable for flexible electronic devices. The connection of these flexible electronic devices through an external circuit is complicated due to their inherent softness and small sizes [5]. Recently, researchers have made some effective attempts to realize the integration of flexible electronic devices. According to the configuration, the integration method of flexible electronic devices can be classified into two categories: all-in-one and assembled. In all-in-one devices, the electrode is imparted for multiple roles, allowing different functions to be realized in a single device. In assembled devices, two individual parts are prefabricated separately and interconnected to realize the integration of different functions.

The use of batteries is indispensable in electronic devices, and the physical and chemical integration of the flexible battery into a wearable electronic system enables complete functions while easing the installation and system scaling. In this chapter, the recent advancements in the

DOI: 10.1201/9781003273677-10

integration of flexible batteries are discussed. In the beginning, a series of integrated devices based on flexible batteries with different functions are presented. For example, solar cell and battery are integrated to realize the energy harvest and storage together, supercapacitor and battery are integrated to achieve high energy density and high power output, and sensor and battery are integrated to form a self-powered system to achieve multifunctionalities. Next, a new type of flexible battery textile that integrates numerous battery units to achieve large-scale application is discussed. Finally, the perspective for the future development of the integration of flexible batteries will be given.

10.2 INTEGRATED SOLAR CELLS AND BATTERIES

Traditionally, energy harvesting and storage methods are developed as independent technologies but are often used together as a power system. For example, a power system based on a solar cell panel and a battery as two independent parts is extensive, inflexible, and complicated. Therefore, integrated energy harvesting and storage devices based on the same electrode are proposed as an effective way to obtain a small-size, flexible, and high-density energy system. A typical integrated device that combines the functions of solar energy conversion and electric energy storage was shown in Figure 10.1 [6]. It consists of a dye-coated TiO_2/WO_3/transparent

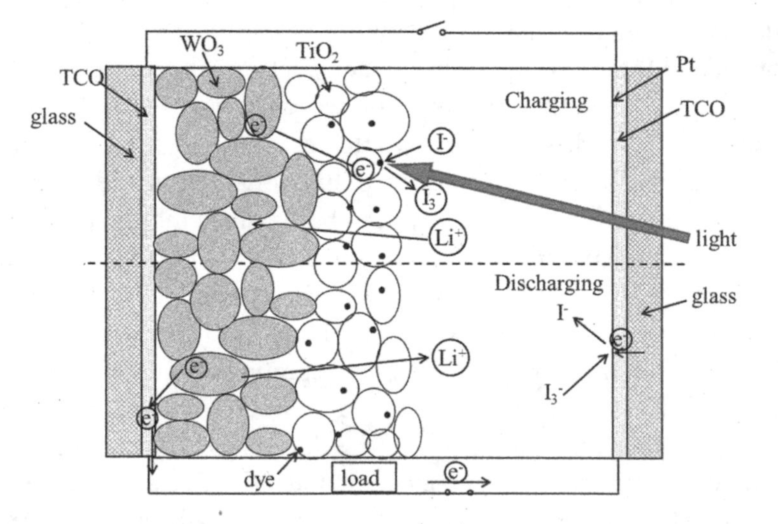

FIGURE 10.1 Schematic diagram of an integrated dye-sensitized solar cell and lithium-ion battery. (Reproduced from Ref. [6] with permission of the Electrochemical Society.)

conductive oxide (TCO) glass working electrode, a Pt/TCO glass opposing electrode, and a LiI/I_2/propylene carbonate/4-*tert*-butyl pyridine electrolyte. When the battery is in open circuit and under illumination, electrons are injected from the photo-excited dyes into the conduction band of TiO_2, which is then diffused into WO_3. To keep the charge balanced, Li^+ was intercalated into the WO_3, leading to the $WO_3|LiWO_3||LiI|LiI_3$ battery. When connected to an external load, i.e., under discharging conditions, electrons are transferred from WO_3 to the Pt electrode. Simultaneously, I_3^- obtained electrons from Pt electrode and reduced to I^-, and the intercalated Li^+ was released. As a result, a charge of 1.8 $C{\cdot}cm^{-2}$ can be stored under irradiation of 1,000 $W{\cdot}m^{-2}$ for 1 h, and the open-circuit voltage in the charge state is 0.6 V. Therefore, both energy conversion and storage can be carried out in one device, which enables high charge storage capacity and simple fabrication.

Based on a similar strategy, an integrated dye-sensitized solar cell and lithium-oxygen battery is also developed by using a redox-coupled dye-sensitized photoelectrode [7]. In a typical nonaqueous rechargeable lithium-oxygen battery, lithium peroxide was formed on the oxygen electrode surface during the discharging process. However, it is challenging to be electrochemically decomposed due to the bulk and insulating property, which leads to a severe charging overpotential issue, e.g., low energy efficiency [8]. The redox shuttle-coupled photoelectrode with the oxygen electrode in the integrated battery system can enable a photo-charging process. During this process, the reduced form of the redox shuttle M^{red} was first photoelectrochemically oxidated to M^{ox}, which subsequently diffused to the oxygen electrode and oxidized the solid lithium peroxide (Figure 10.2a). The charging voltage of the battery equaled the energy difference between the redox potential of the Li^+/Li couple and the quasi-Fermi level of electrons in the TiO_2 photoelectrode, which is close to the conduction band of TiO_2 (Figure 10.2b). By efficiently shuttling charges, the redox shuttle improved lithium peroxide oxidation and reduced the charging overpotential.

The photovoltage generated on the photoelectrode was also utilized to compensate for the charging potential of the lithium-oxygen battery. Based on this photoelectrochemical mechanism, a flexible photo-charging lithium-oxygen battery was recently developed (Figure 10.3a) [9]. A flexible and porous TiN/TiO_2/carbon cloth electrode with both electrocatalytic and photocatalytic activity was synthesized and introduced as a

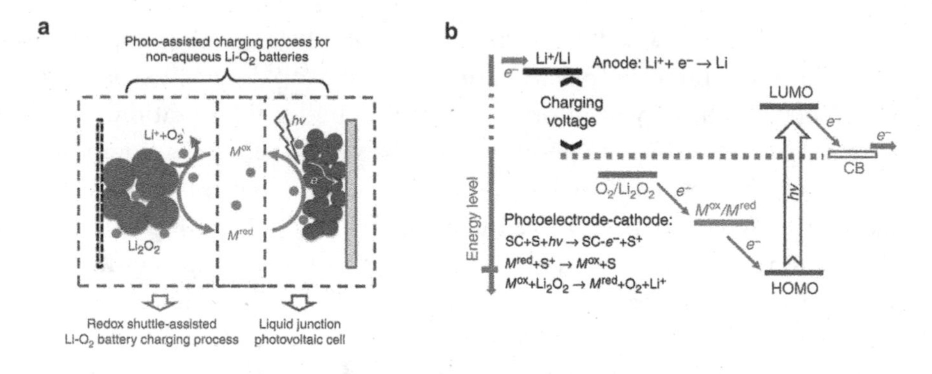

FIGURE 10.2 (a) Schematic diagram of the photoelectrochemical mechanism of a photo-charging lithium-oxygen battery. (b) Energy diagram of a photo-charging lithium-oxygen battery that integrated dye-sensitized semiconductor photoelectrode. (Reproduced from Ref. [7] with permission of Nature Publishing Group.)

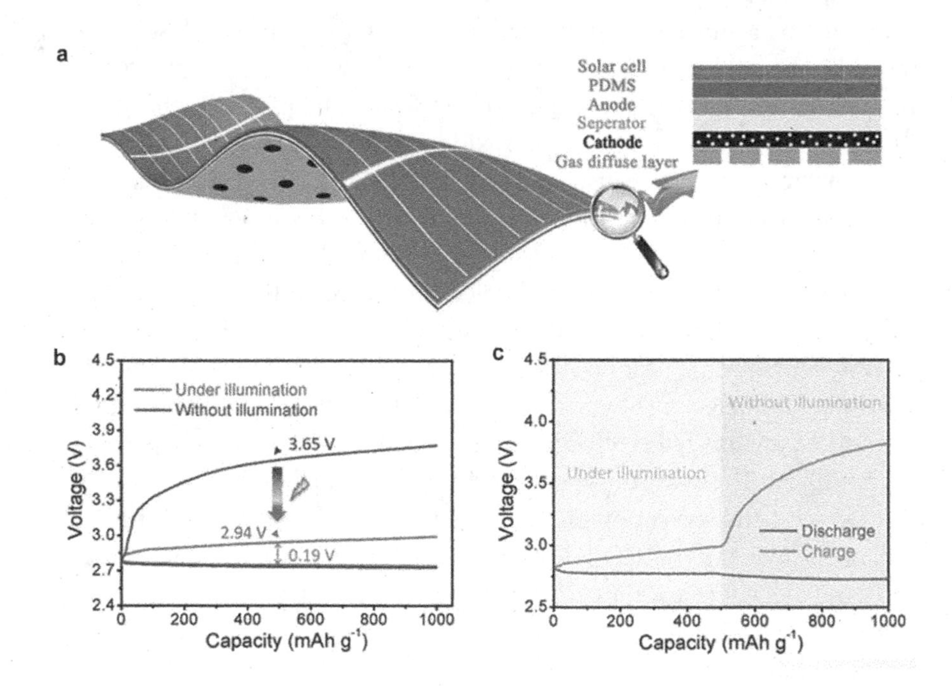

FIGURE 10.3 (a) Schematic diagram of the structure of a flexible integrated self-powered energy system based on photo-charging lithium-oxygen battery. (b) The discharge-charge curves of the flexible photo-charging lithium-oxygen battery with or without illumination. (c) Dynamic light-response discharge-charge curves of the flexible photo-charging lithium-oxygen battery. (Reproduced from Ref. [9] with permission of Wiley-VCH.)

cathode to assemble the battery. With the assistance of photo-charging, the lithium-oxygen battery can achieve an overpotential of 0.19 V with an energy conversion efficiency of ~94% (Figure 10.3b). A dynamic light-response process is also performed, which further confirms that solar energy can be used to compensate the electric energy during the photo-charging process (Figure 10.3c).

Although the above strategy can realize high-density integration to harvest energy and storage devices together, a high concentration redox shuttle is needed, decreasing the battery's stability. To this end, another strategy was proposed by separating energy conversion and storage into two individual parts that were connected [10–13]. Therefore, an energy storage device, e.g., lithium-ion battery, can be integrated instead of the amateur host materials. For instance, an integrated fiber energy device is developed by fabricating a core-sheath structure with lithium-ion storage (LS) part at the core and photoelectric conversion (PC) part at the sheath (Figure 10.4a and b) [13]. For the LS part, carbon nanotube (CNT)/$LiMn_2O_4$ (LMO) and CNT/$Li_4Ti_5O_{12}$ (LTO) composite fibers were used as the positive and negative electrodes, respectively. CNT sheet and spring-shaped Ti/TiO_2 wire were used for the PC part as counter electrode and photoanode, respectively. To fabricate an integrated fiber energy device, the LS part was first inserted into the PC part, and the Ti/TiO_2 wire from one PC part and the CNT sheet from the neighboring PC part were connected through the CNT sheet. Then, the CNT/LMO and CNT/LTO composite fibers of the LS part were connected to the counter electrode and photoanode of the PC part, respectively. A switch was finally incorporated between the PC and LS parts to achieve the photo-charging and discharging processes. The integrated fiber energy device delivered satisfactory photovoltaic conversion and energy storage properties. For the photovoltaic conversion, the high conversion efficiency of 6.05% with an open-circuit voltage of 0.68 V had been achieved. The open-circuit voltage can further be improved to 5.12 V when eight PC units were connected in series (Figure 10.4c). The coulombic efficiency reached over 95% for the energy storage with a specific capacity of ~ 70 $mAh{\cdot}g^{-1}$ and an output voltage of 2.6 V (Figure 10.4d). Due to the higher charging voltage of the lithium-ion battery, several PC units need to be connected in series. As shown in Figure 10.4e, a successful photo-charging process was realized when six PC units were connected, and the photo-charging time had been further reduced for increasing PC units. After fully photo-charged, the

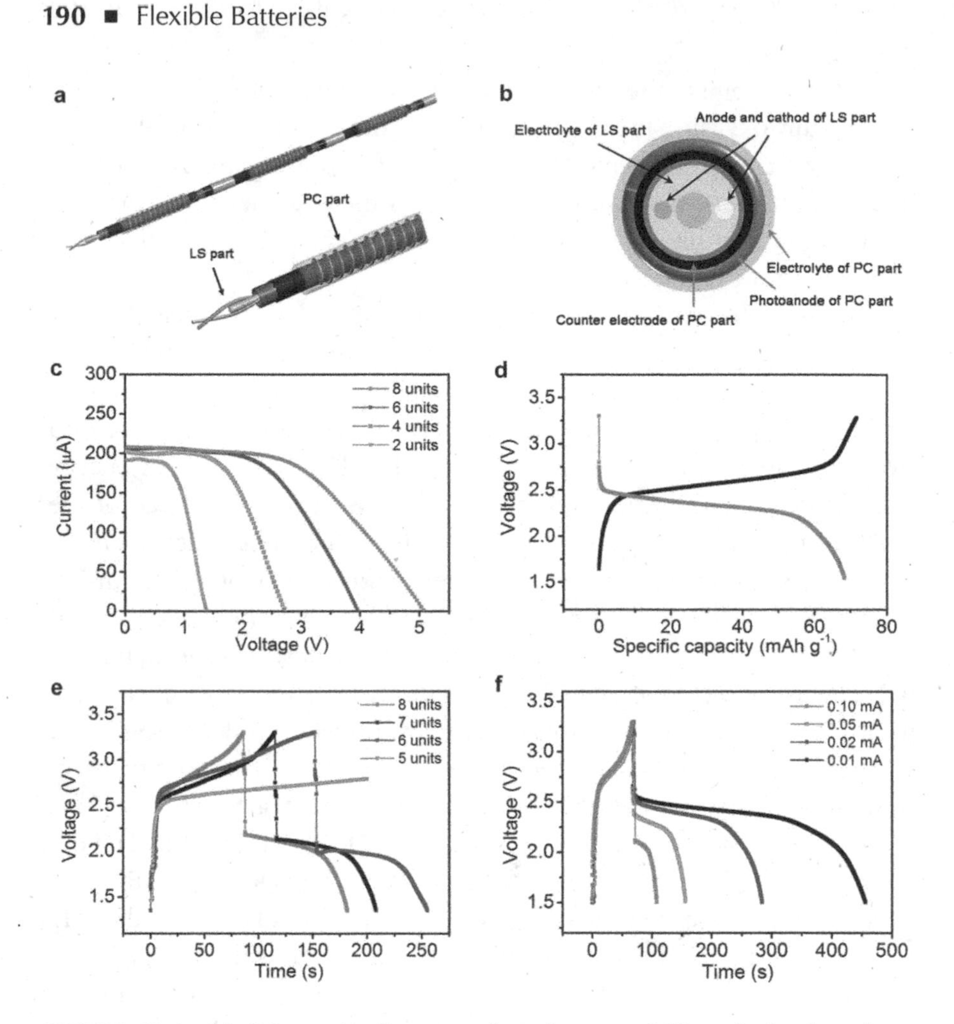

FIGURE 10.4 (a) Schematic diagram of an integrated fiber device based on dye-sensitized solar cell and lithium-ion battery. (b) Cross-sectional views of the integrated fiber device with the PC part at the sheath and the lithium-ion storage (LS) part at the core. (c) *J-V* curves of PC parts with increasing unit numbers. (d) The charge-discharge curves of an LS part at a current of 0.02 mA. (e) Photo-charging and discharging curves with increasing PC units. Discharging current: 0.05 mA. (f) Photo-charging and discharging curves with eight PC units with increasing discharging currents. (Reproduced from Ref. [13] with permission of the Royal Society of Chemistry.)

integrated fiber energy device can discharge within currents from 0.01 to 0.1 mA, and a high energy density of 22 Wh·kg^{-1} had been achieved.

Considering the strong dependence of solar energy on light, a triboelectric nanogenerator (TENG) can be further integrated into a self-powered system to compensate for the energy harvesting efficiency of solar cells [14]. The TENG has been extensively demonstrated to harvest mechanical energy and convert it into electricity based on the coupling effect of triboelectrification and electrostatic induction [15]. As shown in Figure 10.5a, a flexible self-powered device was proposed by integrating Si solar cell, TENG, and lithium-ion battery with a combined thickness of less than 1 mm. Due to the high tensile/impact strength and transmittance, ethylene-*tetra*-fluoro-ethylene (ETFE) can be used as a triboelectrification material for TENG and serve as suitable encapsulating materials for Si

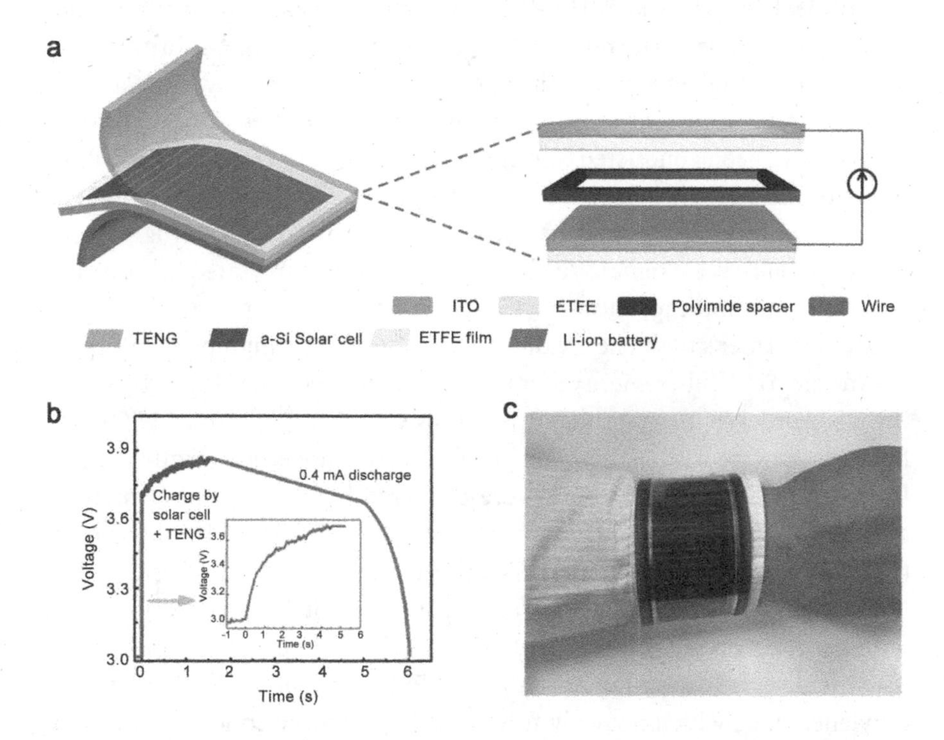

FIGURE 10.5 (a) Schematic diagram of a flexible integrated device based on Si solar cell, triboelectric nanogenerator (TENG), and lithium-ion battery. (b) The charge-discharge curves of the flexible integrated device. (c) Photograph of the flexible integrated device attached to the textile of the wrist. (Reproduced from Ref. [14] with permission of Elsevier.)

solar cells. Therefore, the two parts were prepared as a oneness panel to make the integrated device more compact. Both the solar cell and TENG units can harvest energy to charge the integrated lithium-ion battery unit. As shown in Figure 10.5b, the lithium-ion battery unit can be charged to 3.65 V by the solar cell unit in 6.5 s and then charged to 3.86 V by the TENG unit. After charging, the battery can discharge at 0.4 mA and last for 4.5 h. The integrated device can be attached to clothes on the wrist (Figure 10.5c) or other objects related to human life, e.g., laptop, backpack, and umbrella.

10.3 INTEGRATED SUPERCAPACITORS AND BATTERIES

In practice, both high energy density and high power output are required for an ideal energy system [16–18]. However, it is not easy to simultaneously realize these performances in one device. For instance, lithium-ion batteries have high energy densities but low power outputs, while supercapacitors have high power outputs but low energy densities. Although some studies have been conducted to improve energy density and power output by synthesizing high-performance electrode and electrolyte materials, the effect is limited. How to achieve both high energy density and high power output in one device remains a challenge. Recently, integrated lithium-ion batteries and supercapacitors have become an effective strategy that combines the former's high energy density and the latter's high power output.

An integrated fiber energy storage device had been developed by twisting CNT/ordered mesoporous carbon (OMC), CNT/LTO, and CNT/LMO fiber electrodes with a layer of gel electrolyte together (Figure 10.6a) [19]. The three fiber electrodes were prepared by a co-spinning method that incorporates OMC, LTO, and LMO into CNT fiber. The fiber electrodes showed uniform diameters of 100, 80, and 115 μm, respectively (Figure 10.6b–d). They can be closely intertwined after being coated with gel electrolyte (Figure 10.6e). The gel electrolyte consists of lithium bis(trifluoromethanesulfonyl)imide (LiTFSI), succinonitrile, and poly(-ethyleneoxide), which not only functioned as an ionic conductor but also acted as a separator to prevent short circuits of the device. The integrated fiber energy storage device can act as both a lithium-ion battery and supercapacitor. When the CNT/LTO electrode was paired with the CNT/LMO electrode, the integrated device acted as a lithium-ion battery and exhibited high energy density. The battery segment showed the specific capacity

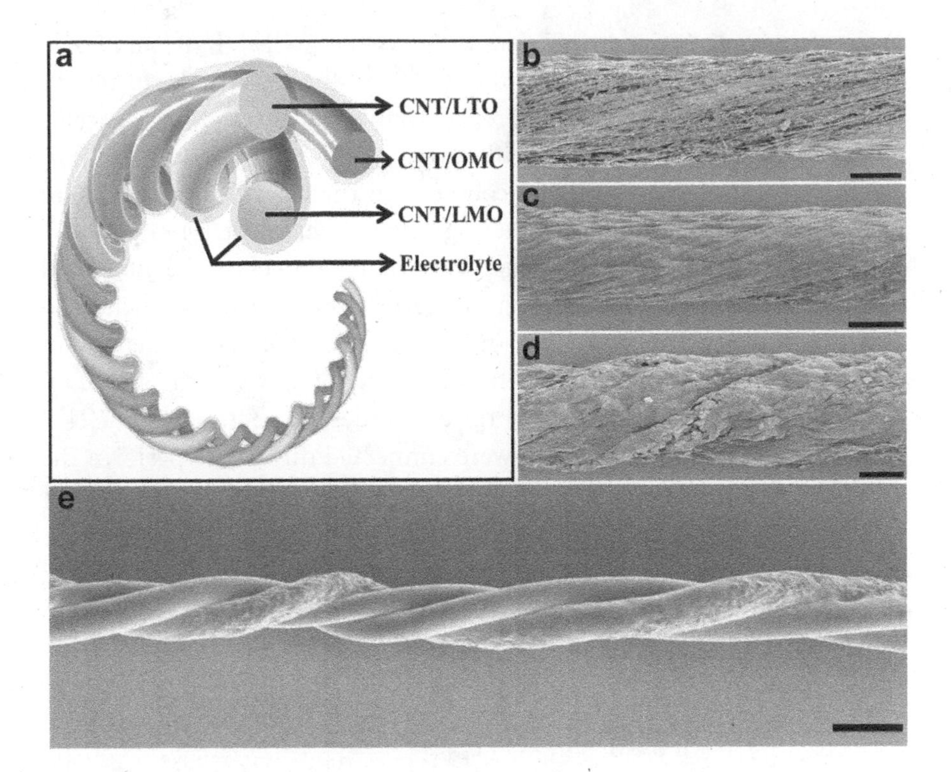

FIGURE 10.6 (a) Schematic diagram of an integrated fiber device based on lithium-ion battery and supercapacitor. (b–d) SEM images of the carbon nanotube (CNT)/ordered mesoporous carbon (OMC), CNT/$Li_4Ti_5O_{12}$ (LTO), and CNT/$LiMn_2O_4$ (LMO) hybrid fibers, respectively. Scale bars: 50 μm. (e) SEM image of the integrated fiber device. Scale bars: 300 μm. (Reproduced from Ref. [19] with permission of Wiley-VCH.)

of 120.5 $mAh \cdot g^{-1}$ and energy density of 98.6 $Wh \cdot kg^{-1}$ with a discharge plateau voltage of ~2.3 V at the current density of 0.5 $mA \cdot g^{-1}$. The integrated device functioned as a supercapacitor when the CNT/LTO electrode was paired with the CNT/OMC electrode. In this case, the supercapacitor segment exhibited a specific capacitance of 22.1 $F \cdot g^{-1}$ at the current density of 1 $A \cdot g^{-1}$ and power density of 4136.7 $W \cdot kg^{-1}$ at the current density of 4 $A \cdot g^{-1}$. The two segments of the integrated device can act together to achieve various power outputs for practical applications. For example, the supercapacitor segment can provide a peak power output to start a flash before photographing, and the lithium-ion battery segment can provide a stable long power output to support the camera's running. In this case, the supercapacitor segment can charge via a self-charge process by the lithium-ion battery segment. As shown in Figure 10.7a, when the CNT/LMO and CNT/OMC electrodes were connected, the electrons flowed from the former to the latter due to the potential difference. In this process, the OMC surface adsorbed many anions (i.e., bis(trifluoromethane)sulfonamide ($TFSI^-$)) in the electrolyte and can therefore couple with the CNT/LTO electrode to form a supercapacitor to show the high power character (Figure 10.7b). Figure 10.7c exhibits a complete charge/discharge process of the integrated energy storage device. The lithium-ion battery segment was first charged and acted as the energy storage unit with a volume energy density of 48.1 $mWh{\cdot}cm^{-3}$ and gravimetric energy density of 85.2 $Wh{\cdot}kg^{-1}$ based on the total volume or mass of the three electrodes, respectively. The CNT/LMO and CNT/OMC electrodes were connected directly to perform the self-charge process for 30 s. After that, the voltage of the supercapacitor

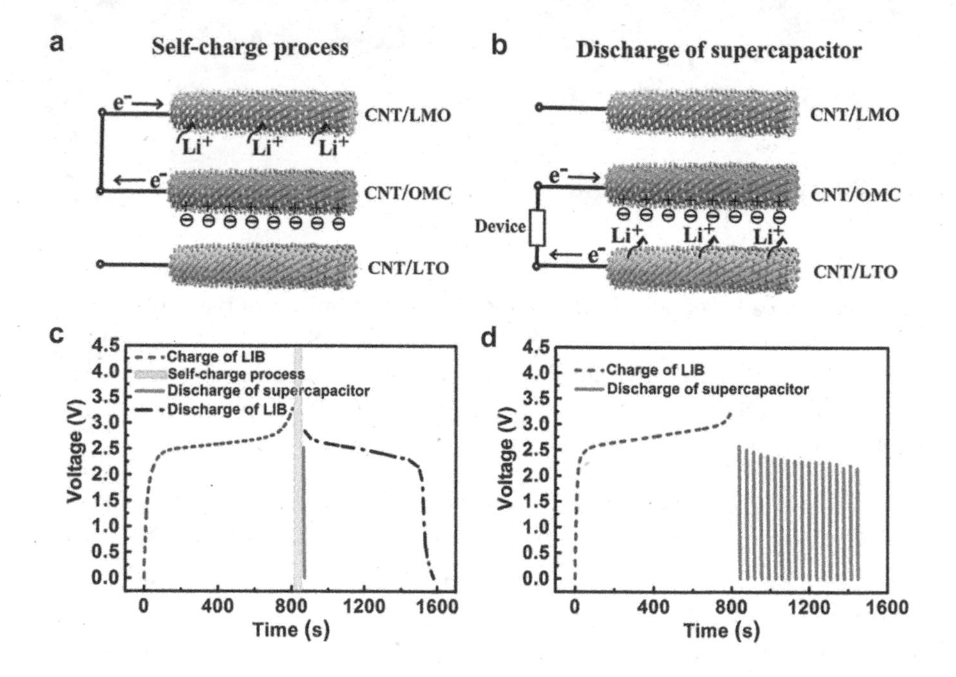

FIGURE 10.7 (a and b) Schematic diagram of the self-charge process and discharge of the supercapacitor segment, respectively. (c) The charge and discharge curves of the integrated device, which was alternately discharged as a lithium-ion battery and a supercapacitor. (d) The charge and discharge behavior of the integrated device in which the lithium-ion battery and supercapacitor segments served as energy storage and output, respectively. (Reproduced from Ref. [19] with permission of Wiley-VCH.)

segment rose to 2.5 V, and it can work as the high power output unit to discharge with a volume power density of 1.07 $W \cdot cm^{-3}$ and gravimetric power density of 5,971.1 $W \cdot kg^{-1}$. Furthermore, the integrated energy storage device can continue to provide the pulse high power output by repeating the self-charge and discharge processes shown in Figure 10.7a and b. As shown in Figure 10.7d, the integrated energy storage device can provide the pulse high power output for 18 times with a total discharge energy density of 31.26 $mWh \cdot cm^{-3}$ and energy efficiency of 65%.

10.4 INTEGRATED SENSORS AND BATTERIES

Wearable sensing technologies can efficiently monitor health conditions by detecting physiological signals in real time, gaining increasing attention in biomedical and healthcare fields [20]. However, most sensors need electricity to work, which makes it necessary to equip them with an additional power supply, increasing the system's complexity. To this end, the effective integration of a flexible battery with a wearable sensor can realize a self-powered system to achieve multifunctionalities for potential applications [21]. In this section, the integrated self-powered system based on sensor and battery is discussed.

Figure 10.8a shows the device structure and operational principles of a typical self-powered sensor system [22]. The device was fabricated based on paper fluidics and used a Prussian blue spot electro-deposited on a transparent indium-doped tin oxide (ITO) thin film as an indicator of the sensor unit. The paper substrate was preloaded with a glucose oxidase and $Fe(CN)_6^{3-}$ for glucose sensor and horseradish peroxidase and $Fe(CN)_6^{4-}$ for the H_2O_2 sensor. When glucose in artificial urine was adsorbed on the reaction zone, glucose oxidase catalyzed glucose oxidation with simultaneous reduction of $Fe(CN)_6^{3-}$ to $Fe(CN)_6^{4-}$ and activated Prussian blue to colorless Prussian white. A similar principle was used to detect H_2O_2, but the indicator converted from a colorless Prussian white to a Prussian blue. The sensor unit was powered by an aluminum-air battery assembled from a piece of aluminum foil as the anode, high surface area activated carbon as the cathode, and paper as the separator. The same ITO electrodes used for the sensor unit were used as the current collectors for the Al-air battery. In order to eliminate the self-discharge during long-term storage, artificial urine was used as the battery electrolyte to activate the battery before sensing. As shown in Figure 10.8b, the average open-circuit voltage of the Al-air battery was ~0.94 V with a maximum short circuit current

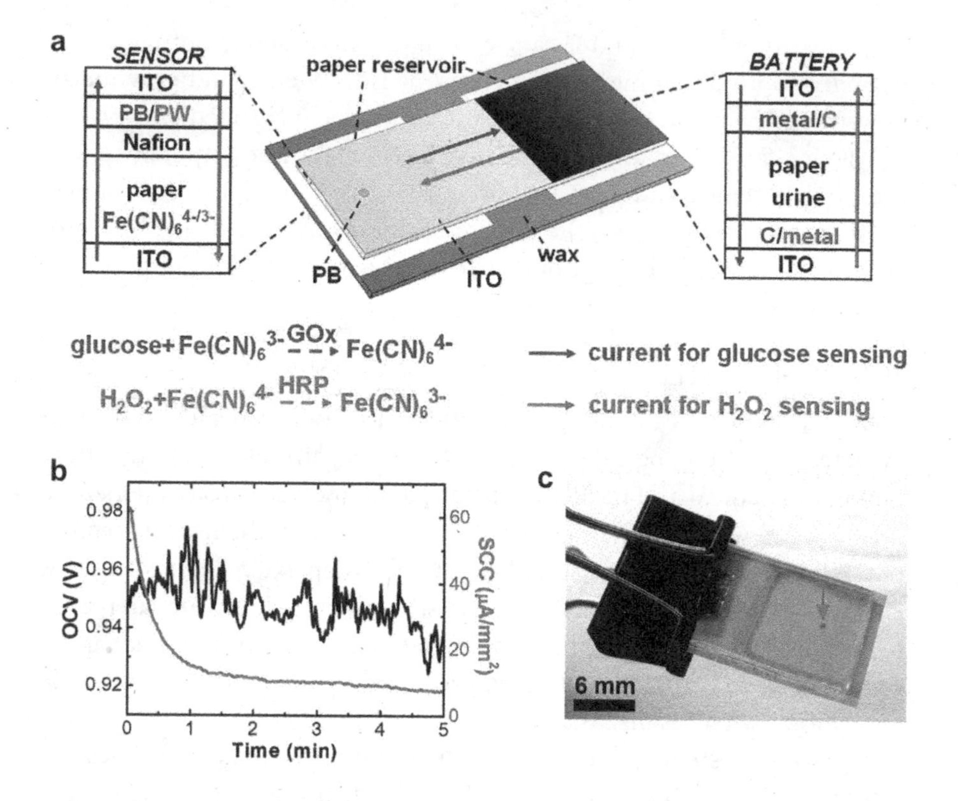

FIGURE 10.8 (a) Schematic diagram of the operational principle of the integrated paper device based on electrochemical sensor and Al-air battery. (b) Open-circuit voltage and short circuit current measurements for the Al-air battery. (c) Photograph of the integrated paper device used for glucose detection. The red arrow indicates the sensor region, which can change from blue to colorless in the presence of glucose. (Reproduced from Ref. [22] with permission of the American Chemical Society.)

of 60 $\mu A \cdot mm^{-2}$. Figure 10.8c shows the complete integrated paper device, and the red arrow indicated the glucose sensor region, which can change from blue to colorless in the presence of artificial urine contained glucose. Although the integrated paper device was convenient with electrochromic read-out, it had some limitations. On the one hand, its detection accuracy is limited, unable to detect the exact content of the target analyte. On the other hand, it is difficult to combine with portable electronic equipment, e.g., mobile phones, to realize digital diagnosis and response.

Compared with colorimetric sensors, a sensor with electrical signal response can provide higher accuracy and can be easily integrated with other electronic components to achieve the true potential of real-time data acquisition [23]. For instance, an integrated cloth device based on electrochemical fabric and fiber lithium-ion battery has been developed for *in situ* perspiration analysis (Figure 10.9a) [24] The electrochemical fabric was constructed by weaving different sensing fibers that could detect glucose, Na^{+}, K^{+}, Ca^{2+}, and pH as the building blocks. All the sensing fibers are prepared *via* a general strategy by depositing different active materials (Prussian blue and glucose oxidase for glucose sensing fiber, poly(3, 4-ethylenedioxythiophene): polystyrene sulfonate and ion-selective ionophore for ion-sensing fiber, and polyaniline for pH sensing fiber) on CNT fiber electrode. The fiber lithium-ion battery is fabricated by twisting a $LiCoO_2$/stainless steel wire and graphite/Cu wire electrodes and then inserting them into a heat-shrinkable tube with electrolyte LB303. For the *in situ* analysis of sweat, the sensing fibers and fiber lithium-ion battery were integrated into the garment. Two flexible integrated chips were used to wirelessly transfer data to a Bluetooth-enabled smartphone, with a custom-developed application downloaded. As shown in Figure 10.9b–d, the integrated cloth device can obtain a real-time analysis of sweat, which is consistent with the *ex situ* measurements.

The textile TENG can be further integrated with a flexible lithium-ion battery belt and sensor to form a self-charging power integrated system

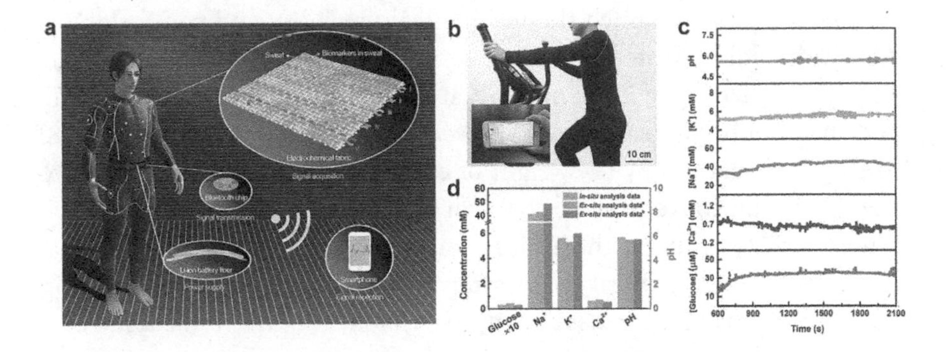

FIGURE 10.9 (a) Schematic diagram of an integrated cloth device based on the lithium-ion battery and electrochemical sensors. (b) Photograph of a subject wearing the integrated cloth device while running. (c) Real-time sweat analysis using the integrated cloth device. (d) Comparison of *ex situ* data from the collected sweat samples with that *in situ*. [a]Measured by using sensing fibers. [b]Measured by using commercial instruments. (Reproduced from Ref. [24] with permission of Wiley-VCH.)

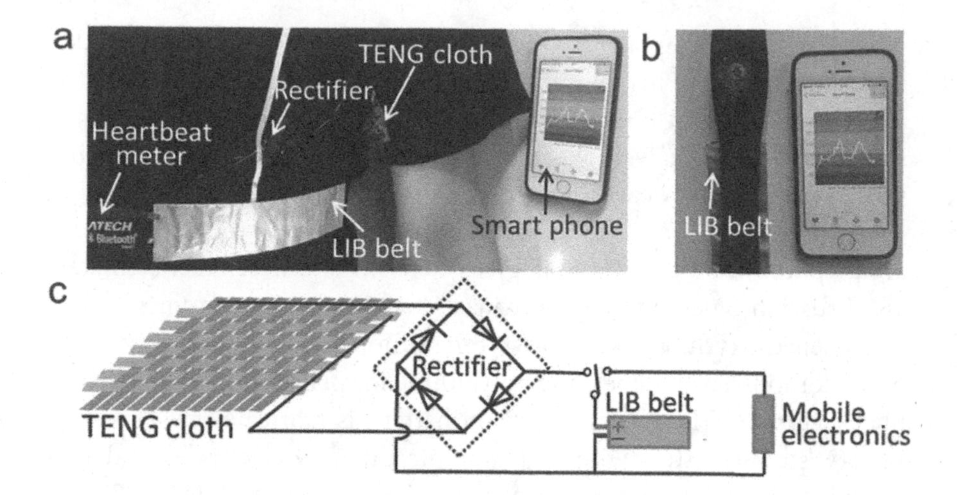

FIGURE 10.10 (a) Photograph of a flexible integrated device based on triboelectric nanogenerator (TENG), lithium-ion battery, and heartbeat sensor, which has remote communication with a smartphone. (b) The rear side of the lithium-ion battery and heartbeat sensor. (c) The equivalent electrical circuit of the flexible integrated device. (Reproduced from Ref. [25] with permission of Wiley-VCH.)

(Figure 10.10a) [25]. In this case, a common flexible polyester fabric coated with conductive Ni film (Ni-cloth) was used as an electrode of the textile TENG and current collector of the lithium-ion battery belt. Therefore, the integrated system retained the mechanical flexibility and comfortability of the original polyester cloth. The textile TENG exhibited a maximum peak power density of 393.7 $mW{\cdot}m^{-2}$ at an external resistance of ~70 MΩ, and the lithium-ion battery belt showed a stable discharge capacity of 81 $mAh{\cdot}g^{-1}$ normalized by the weight of cathode. After integration, the textile TENG can convert the mechanical energy of various human motions into electricity to charge the lithium-ion battery belt, ensuring the regular operation of the heartbeat meter strap to obtain the physiological information of human and transmit to a smartphone (Figure 10.10b and c).

10.5 BATTERY TEXTILES

Flexible batteries usually have a low energy density, so they need to be integrated into a battery pack to meet the requirements of practical applications. Although thin-film batteries are flexible, their flexibility will significantly reduce when many of them are integrated. Moreover, they are neither breathable nor comfortable, whereas breathability and comfort are highly desired for wearable applications. Compared with thin-film batteries, battery textiles can bear various complex deformations, e.g., bending and twisting, and they also provide an effective and comfortable interface with our body [26–28]. Therefore, the battery textile would allow long-term use without discomfort for users and can be integrated at a large scale for practical applications. Recently, there are two main strategies for the construction of battery textiles. One is to fabricate flexible battery modules on the textile substrate directly and further integrate them. The other is to integrate the fiber batteries into the textile by weaving.

Figure 10.11a shows a flexible battery textile module fabricated by assembling Ni-coated polyester fabric/LTO, Ni-coated polyester fabric/LFP electrodes, and separator in a conventional stacking configuration [29]. The 3D fiber networks that inherently existed within the textiles enabled the battery to be flexible and integrated with other flexible electronic devices, such as the rollable display. Notably, the module design of the battery textile offered additional degrees of freedom in the connection. For instance, a connection in parallel could increase the overall capacity, whereas a connection in series could enhance the operation voltage. As shown in Figure 10.11b, each battery textile module delivered an output voltage of ~1.8 V and the entire battery textile consisting of 16 modules thus delivered ~28.8 V. The flexible, foldable, and large-scale characteristics of the module design allowed the battery textile to be attached on the outdoor tent or roller blind. They can thus be folded/rolled when the main body of the outdoor tent or roller blind was folded/rolled (Figure 10.11c and d). It could be further combined with flexible solar cells to achieve efficient energy harvesting and storage.

Compared with the first strategy, the battery textile constructed by the second strategy can be more flexible and breathable. Inspired by traditional textile technology, the prepared fiber batteries can be further integrated into the textile by weaving. As shown in Figure 10.12a, fiber batteries and cotton fibers can be woven into a textile together to form a battery textile [29]. The battery textile exhibited an interleaved structure and can thus bear various deformations such as bending, folding, and twisting (Figure 10.12b and c). After fabrication of a wristband and wear on the wrist, the battery textile can effectively power a commercial LED screen (Figure 10.12d). In this case, a single fiber battery can be considered a module unit while the output voltage and capacitor of the battery textile can be adjusted by changing their number and connection.

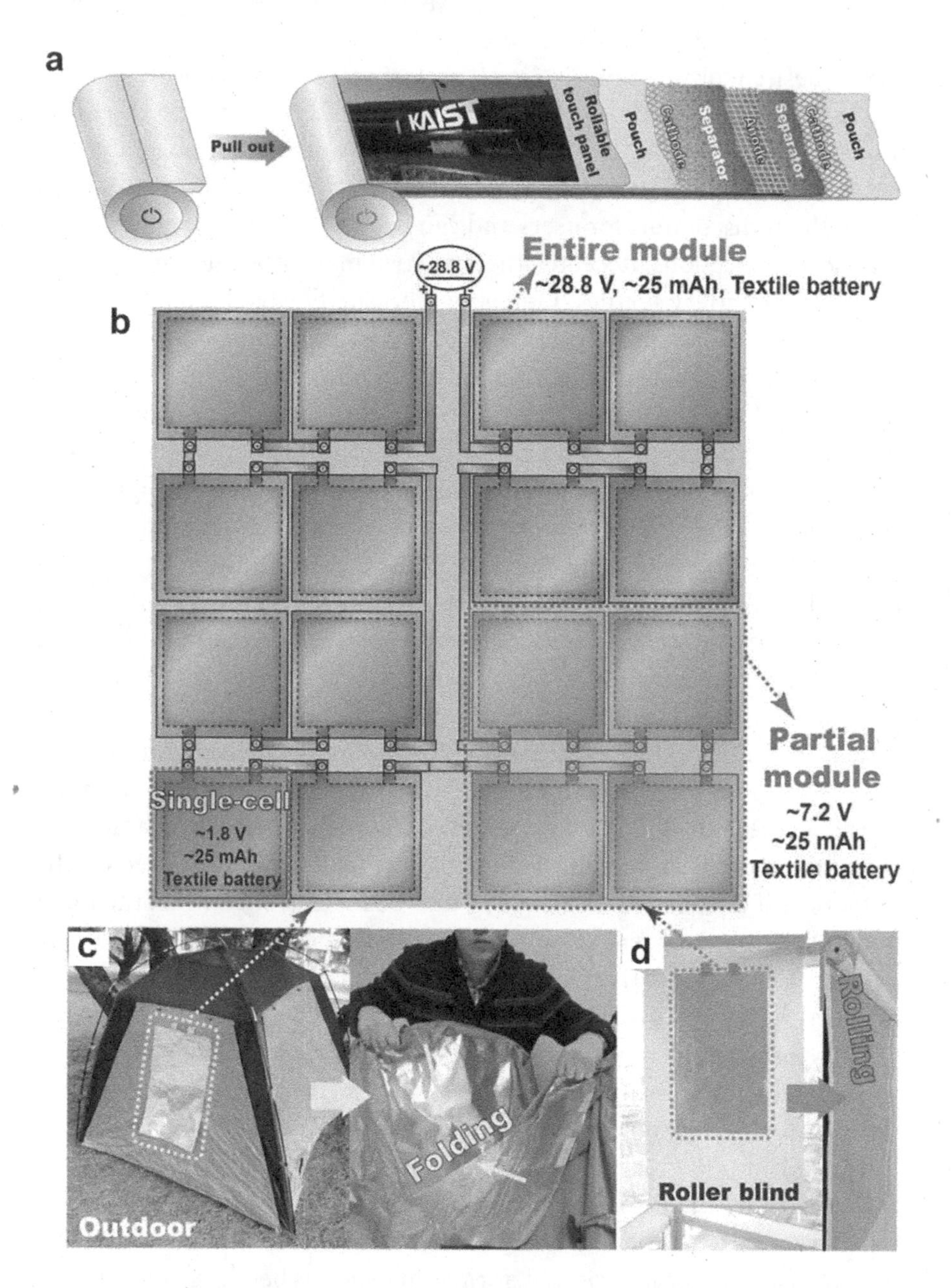

FIGURE 10.11 (a) Schematic diagram of a flexible display integrated with a battery textile. (b) Schematic diagram of the battery textile module. (c and d) Various potential applications based on battery textile. Photographs of a battery textile were attached to (c) an outdoor tent and (d) a roller blind on a building window. (Reproduced from Ref. [28] with permission of the Royal Society of Chemistry.)

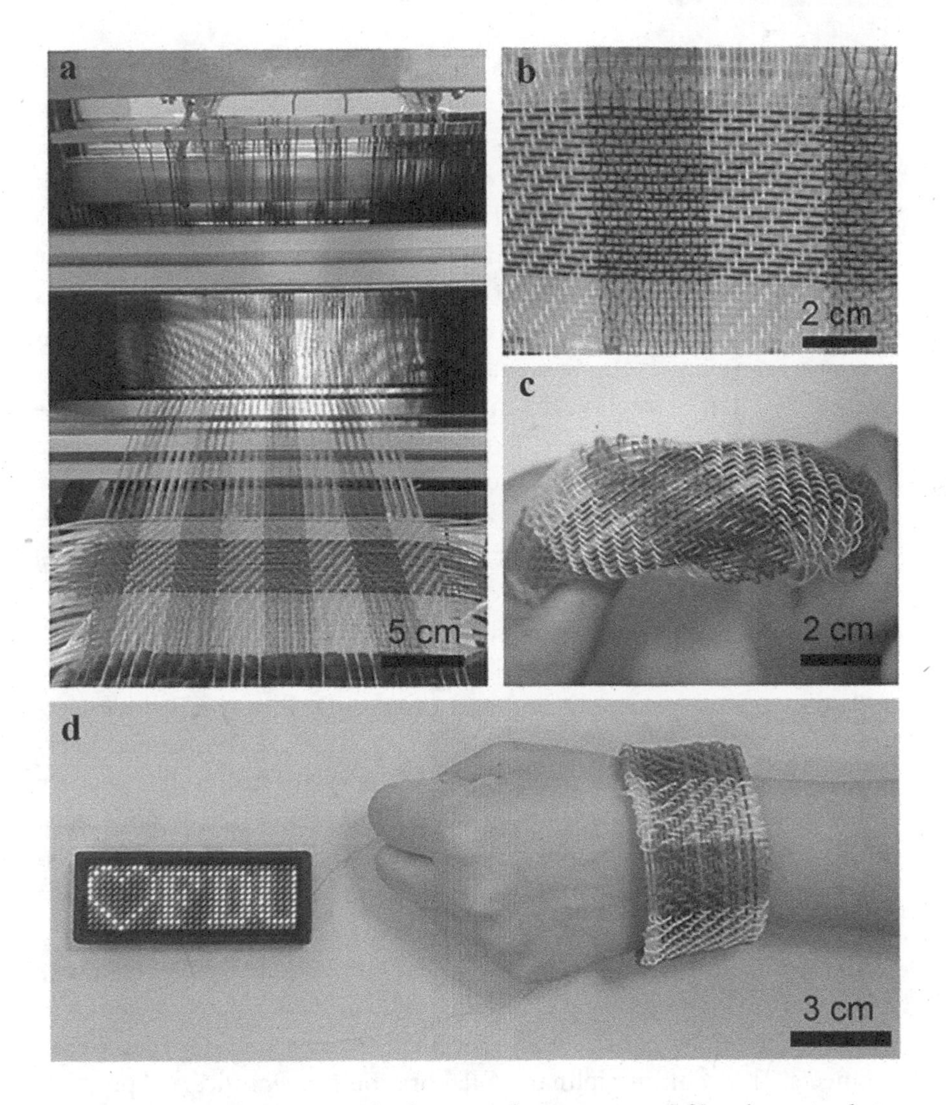

FIGURE 10.12 (a) Photograph showing the progress of fiber batteries being woven into a textile. (b) Photographs of the battery textile before and after twisting. (c) The battery textile woven on the wrist to power a light-emitting diode display screen. (Reproduced from Ref. [29] with permission of Wiley-VCH.)

10.6 PERSPECTIVE

In the past few years, the emerging flexible batteries and integrated devices have become the most promising branch of modern electronics and have aroused increasing research and commercial interest. However, there are still many challenges and technological gaps hindering their practical applications.

First, the comprehensive performance of the flexible battery integrated devices needs to be improved. For example, the overall energy conversion efficiency of an integrated solar cell and lithium-ion battery was only ~10% [30]. It is well recognized that improving the power conversion efficiency of solar cell units or the coulombic efficiency of lithium-ion battery units is an efficient strategy to improve the overall energy conversion efficiency. Therefore, developing the integrated unit with high performance separately is essential to improve the integrated devices. In addition, optimizing the structure and connection of different integrated units to reduce the loss of electrical energy in the electrode is another way to enhance comprehensive performance.

Second, the complete function of an integrated electronic system requires the cooperation of numerous electronic components. In addition to the battery, solar cell, supercapacitor, and sensor, other electronic components, e.g., transistor and memory, are essential because they can perform electronic functions such as data recording and processing. Ideally, all these electronic components of a wearable electronic system are flexible and lightweight. However, flexible transistors and memory are still in their infancy and are far from practical applications [31–33]. Therefore, some commercial silicon-based electronic components are used in the wearable electronic system to realize the electronic functions at the present stage. The flexibility and integration among all components need to be further improved.

The application of flexible batteries and their integrated devices requires professionals' joint efforts in many different fields, such as material scientists, electrochemical scientists, electrical engineers, and mechatronics engineers. This multidisciplinary collaboration can significantly promote the process of its practical application, which will provide new opportunities to improve our future life.

REFERENCES

1. Liu, Y. He, K. Chen, G. Leow, W. R. Chen, X. 2017. Nature-inspired structural materials for flexible electronic devices. *Chemical Reviews* 117: 12893–12941.
2. Gao, W. Emaminejad, S. Nyein, H. Y. Y. Challa, S. Chen, K. Peck, A. Fahad, H. M. Ota, H. Shiraki, H. Kiriya, D. 2016. Fully integrated wearable sensor arrays for multiplexed in situ perspiration analysis. *Nature* 529: 509–514.

3. Yang, Y. Gao, W. 2019. Wearable and flexible electronics for continuous molecular monitoring. *Chemical Society Reviews* 48: 1465–1491.
4. Wang, L. Fu, X. He, J. Shi, X. Chen, T. Chen, P. Wang, B. Peng, H. 2020. Application challenges in fiber and textile electronics. *Advanced Materials* 32: 1901971.
5. Huang, S. Liu, Y. Zhao, Y. Ren, Z. Guo, C. F. 2019. Flexible electronics: stretchable electrodes and their future. *Advanced Functional Materials* 29: 1805924.
6. Hauch, A. Georg, A. Krašovec, U. O. Orel, B. 2002. Photovoltaically self-charging battery. *Journal of the Electrochemical Society* 149: A1208.
7. Yu, M. Ren, X. Ma, L. Wu, Y. 2014. Integrating a redox-coupled dye-sensitized photoelectrode into a lithium-oxygen battery for photoassisted charging. *Nature Communications* 5: 1–6.
8. Wang, L. Pan, J. Zhang, Y. Cheng, X. Liu, L. Peng, H. 2018. A Li-air battery with ultralong cycle life in ambient air. *Advanced Materials* 30: 1704378.
9. Yang, X. Feng, X. Jin, X. Shao, M. Yan, B. Yan, J. Zhang, Y. Zhang, X. 2019. An illumination-assisted flexible self-powered energy system based on a Li-O_2 battery. *Angewandte Chemie International Edition* 58: 16411–16415.
10. Gurung, A. Reza, K. M. Mabrouk, S. Bahrami, B. Pathak, R. Lamsal, B. S. Rahman, S. I. Ghimire, N. Bobba, R. S. Chen, K. 2020. Rear-illuminated perovskite photorechargeable lithium battery. *Advanced Functional Materials* 30: 2001865.
11. Gurung, A. Chen, K. Khan, R. Abdulkarim, S. S. Varnekar, G. Pathak, R. Naderi, R. Qiao, Q. 2017. Highly efficient perovskite solar cell photocharging of lithium ion battery using DC–DC booster. *Advanced Energy Materials* 7: 1602105.
12. Agbo, S. N. Merdzhanova, T. Yu, S. Tempel, H. Kungl, H. Eichel, R. A. Rau, U. Astakhov, O. 2016. Development towards cell-to-cell monolithic integration of a thin-film solar cell and lithium-ion accumulator. *Journal of Power Sources* 327: 340–344.
13. Sun, H. Jiang, Y. Xie, S. Zhang, Y. Ren, J. Ali, A. Doo, S. G. Son, I. H. Huang, X. Peng, H. 2016. Integrating photovoltaic conversion and lithium ion storage into a flexible fiber. *Journal of Materials Chemistry A* 4: 7601–7605.
14. Ma, W. Li, X. Lu, H. Zhang, M. Yang, X. Zhang, T. Wu, L. Cao, G. Song, W. 2019. A flexible self-charged power panel for harvesting and storing solar and mechanical energy. *Nano Energy* 65: 104082.
15. Zhu, G. Peng, B. Chen, J. Jing, Q. Wang, Z. L. 2015. Triboelectric nanogenerators as a new energy technology: from fundamentals, devices, to applications. *Nano Energy* 14: 126–138.
16. Fan, Z. Yan, J. Wei, T. Zhi, L. Ning, G. Li, T. Wei, F. 2011. Asymmetric supercapacitors based on graphene/MnO_2 and activated carbon nanofiber electrodes with high power and energy density. *Advanced Functional Materials* 21: 2366–2375.

17. Li, B. Dai, F. Xiao, Q. Yang, L. Shen, J. Zhang, C. Cai, M. 2016. Nitrogen-doped activated carbon for a high energy hybrid supercapacitor. *Energy & Environmental Science* 9: 102–106.
18. Qu, G. Cheng, J. Li, X. Yuan, D. Chen, P. Chen, X. Wang, B. Peng, H. 2016. A fiber supercapacitor with high energy density based on hollow graphene/conducting polymer fiber electrode. *Advanced Materials* 28: 3646–3652.
19. Zhang, Y. Zhao, Y. Cheng, X. Weng, W. Ren, J. Fang, X. Jiang, Y. Chen, P. Zhang, Z. Wang, Y. 2015. Realizing both high energy and high power densities by twisting three carbon-nanotube-based hybrid fibers. *Angewandte Chemie International Edition* 54: 11177–11182.
20. Zhu, Z. Li, R. Pan, T. 2018. Imperceptible epidermal-iontronic interface for wearable sensing. *Advanced Materials* 30: 1705122.
21. Zhai, Q. Xiang, F. Cheng, F. Sun, Y. Yang, X. Lu, W. Dai, L. 2020. Recent advances in flexible/stretchable batteries and integrated devices. *Energy Storage Materials* 33: 116–138.
22. Liu, H. Crooks, R. M. 2012. Based electrochemical sensing platform with integral battery and electrochromic read-out. *Analytical Chemistry* 84: 2528–2532.
23. Asadian, E. Ghalkhani, M. Shahrokhian, S. 2019. Electrochemical sensing based on carbon nanoparticles: A review. *Sensors and Actuators B: Chemical* 293: 183–209.
24. Wang, L. Wang, L. Zhang, Y. Pan, J. Li, S. Sun, X. Zhang, B. Peng, H. 2018. Weaving sensing fibers into electrochemical fabric for real-time health monitoring. *Advanced Functional Materials* 28: 1804456.
25. Pu, X. Li, L. Song, H. Du, C. Zhao, Z. Jiang, C. Cao, G. Hu, W. Wang, Z. L. 2015. A self-charging power unit by integration of a textile triboelectric nanogenerator and a flexible lithium-ion battery for wearable electronics. *Advanced Materials* 27: 2472–2478.
26. Zhang, Y. Zhao, Y. Ren, J. Weng, W. Peng, H. 2016. Advances in wearable fiber-shaped lithium-ion batteries. *Advanced Materials* 28: 4524–4531.
27. Gulzar, U. Goriparti, S. Miele, E. Li, T. Maidecchi, G. Toma, A. De Angelis, F. Capiglia, C. Zaccaria, R. P. 2016. Next-generation textiles: from embedded supercapacitors to lithium ion batteries. *Journal of Materials Chemistry A* 4: 16771–16800.
28. Kim, J. S. Lee, Y. H. Lee, I. Kim, T. S. Ryou, M. H. Choi, J. W. 2014. Large area multi-stacked lithium-ion batteries for flexible and rollable applications. *Journal of Materials Chemistry A* 2: 10862–10868.
29. Zhang, Y. Jiao, Y. Lu, L. Wang, L. Chen, T. Peng, H. 2017. An ultraflexible silicon–oxygen battery fiber with high energy density. *Angewandte Chemie International Edition* 56: 13741–13746.
30. Fu, X. Sun, H. Xie, S. Zhang, J. Pan, Z. Liao, M. Xu, L. Li, Z. Wang, B. Sun, X. 2018. A fiber-shaped solar cell showing a record power conversion efficiency of 10%. *Journal of Materials Chemistry A* 6: 45–51.
31. Ni, Y. Wang, Y. Xu, W. 2020. Recent process of flexible transistor-structured memory. *Small* 1905332.

32. Owyeung, R. E. Terse-Thakoor, T. Rezaei Nejad, H. Panzer, M. J. Sonkusale, S. R. 2019. Highly flexible transistor threads for all-thread based integrated circuits and multiplexed diagnostics. *ACS Applied Materials & Interfaces* 11: 31096–31104.
33. Xu, X. Zhou, X. Wang, T. Shi, X. Liu, Y. Zuo, Y. Xu, L. Wang, M. Hu, X. Yang, X. 2020. Robust DNA-bridged memristor for textile chips. *Angewandte Chemie International Edition* 59: 12762–12768.

CHAPTER 11

Summary and Outlook

SINCE THE FLOURISHING OF portable and wearable devices, flexible batteries with both high electrochemical properties and flexibility are desired. Flexible thin-film batteries are first developed by simply transforming the bulky and rigid electrodes into flexible electrodes. Solid-state electrolytes such as hydrogel electrolytes are also fabricated and sandwiched by the flexible electrodes, simultaneously serving as electrolytes and separators. Despite their advancement in flexibility to a certain degree, their two-dimensional nature still restricts their breathability and integration capability. Therefore, flexible fiber batteries are evolved with the improvement of fiber electrodes. In this book, we particularly emphasize fiber batteries based on aligned carbon nanotubes (CNTs). The CNT fiber possesses merits of high electrical conductivity, high mechanical strength, and light weight, ensuring their dominating position in fiber batteries. Herein, in the last chapter, our discussions on flexible batteries will end up in concluding their advantages, applications, challenges, and future directions.

11.1 ADVANTAGES

Since their realization, flexible batteries have been endowed with advantages over their bulky counterparts, featuring the requirements from the development of electronic devices. The advantages of flexible batteries are concluded as follows.

DOI: 10.1201/9781003273677-11

11.1.1 Flexibility

The flexible components, especially flexible electrodes, facilitate the flexible batteries to work normally under various deformations, such as bending, twisting, and stretching. Therefore, more application scenarios are enlightened in portable, wearable, and even implantable devices. The flexibility enables sustainable functioning without evident deterioration in electrochemical performance.

11.1.2 Miniaturization

From the initial computer of several tons to modern computers in millimeters, miniaturization has become the trend of modern electronics. Therefore, thinner and lighter batteries are desired as power sources for such miniatured devices with excellent performances. However, traditional bulky and rigid batteries cannot meet this requirement due to fabrication difficulties. Flexible batteries with planar or fibrous configurations are easily miniatured due to the adjustability of electrode materials and solid electrolytes. For example, the fiber electrodes used for fiber batteries are generally prepared with diameters of several micrometers. Hence, by precisely controlling the fabrication technique, the overall volume of the as-fabricated full battery can be less than 0.1 mm. Up to now, flexible batteries have been proven to power light-emitting diodes in light-assisted dental therapy when embedded in dental braces.

11.1.3 Wearability

Compared with conventional bulky batteries, flexible batteries are more favorable for application in wearable electronics. The requirement for wearable devices features mainly three aspects: (1) the compatibility and conformability to the human body; (2) the light weight to reduce the extra burden to the body; and (3) the breathability for better wearing experience. In consideration of these features, power textiles woven from fiber batteries exhibit more advantages over thin-film batteries due to higher flexibility, lighter weight, and better breathability. So far, significant progress has been made to integrate fiber batteries into multifunctional textiles.

11.1.4 Other Advantages

In addition to the advantages mentioned above, flexible batteries are provided with unique advantages based on their working mechanisms and structural configurations. For example, conventional batteries are generally fabricated in a parallel structure. In comparison, fiber batteries can be fabricated into a coaxial structure, exhibiting better structural integrity and uniform diffusion behavior. The integration of flexible batteries is also easier in comparison with conventional bulky batteries. The flexible electrodes can be directly connected with other devices without additional wiring. Especially for fiber flexible batteries, the integration can be realized during the weaving process.

11.2 APPLICATIONS

The application of flexible batteries has been anticipated long ago and realized lately with an encouraging application potential.

11.2.1 Portable Devices

Flexible batteries provide possibilities for replacing conventional bulky battery modules in portable devices, such as cameras, mobile phones, and laptops. Recent years have witnessed the prospering of wearable electronics, such as Galaxy Gear by Samsung Co. Ltd. and Apple Watch by Apple Inc. The flexible and lightweight batteries can be directly woven into the structural components of such wearable devices, such as the watchband of a watch, thus significantly reducing the volume and improving the compatibility with the human body. The integrated flexible battery with energy harvesting devices or supercapacitors represents a promising direction in the future.

11.2.2 Miniature Devices

Besides flexibility, miniaturization is another mainstream for current electronic devices. For example, micromotors have been regarded as promising actuators for drug delivery, cancer diagnosis, and noninvasive microsurgery, with sizes of 2–4 mm and weights of ~0.2 g. To make the best use of the micromotors, the size of power source must be taken into consideration. Flexible batteries can be easily miniaturized, so they are ideal candidates for driving micromotors and micro and nanodevices.

11.2.3 Aerospace Applications

Flexible batteries are also drawing attention from aerospace applications in the future, as flexible batteries share advantages of flexibility, light weight, ease for integration, and high efficiency. Nowadays, although it is fine for an astronaut to work at the space or on moon, it is almost impossible for ordinary people to travel or even live there based on the current rigid equipment including batteries. To this end, flexible batteries are a necessity for the real exploration of the human kind in the following 50 or 100 years. For instance, flexible batteries are integrated into a spacesuit that should also be flexible to make daily activities possible at the space or on the other star. Of course, it is still far away from the real travel of many people, but basic studies and some application models should be made as soon as possible.

11.2.4 Wearable Devices

The most appealing application of flexible batteries lies in the newly emerging wearable electronics. With the growing need for real-time communication and health monitoring, wearable devices such as smart bracelets and watches are developed, but the rigid batteries cannot satisfy the urging needs for miniaturization and conformability to the human body. Therefore, a viable strategy is to incorporate flexible batteries into electronic systems as building components. Adopting the superior flexibility, light weight, and integration capability of flexible batteries, flexible electronics with smaller sizes, better flexibility, and more functions can be realized.

11.3 CHALLENGES AND FUTURE DIRECTIONS

Owing to the unique advantages of flexible batteries, the application scenarios have been enlightened in the past decades. However, the future development of flexible batteries is still hindered by some unignorable challenges and should be addressed in future optimizations.

11.3.1 Flexible Electrodes

Flexible electrodes play critical roles in both flexibility and electrochemical performances of flexible batteries. On the one hand, flexible electrodes and their interfaces with electrolytes typically restrict their flexibility. On the other hand, flexible electrodes' conductivity, surface area, and pseudocapacity contribute to the overall electrochemical performance. However, the present flexible electrodes cannot fully meet these requirements. The initially designed flexible electrodes are metal foils or wires with high conductivity of 10^5 S·cm^{-1}, but their flexibility is inferior, along with small specific surface area and low loading capacity. Carbonaceous materials are superior in flexibility and loading capability with comparatively lower conductivity of 10^2–10^3 S·cm^{-1}. To enable both better flexibility and electrochemical performance, it is crucial to develop novel flexible electrodes to combine the advantages of various available materials with high performance, pointing out a future direction for development.

11.3.2 Electrochemical Performance

Despite the flexibility provided by flexible batteries, their electrochemical performance is compromised due to the modified electrolyte and configuration compared with their bulky counterparts. However, it is well known that to realize the application of flexible batteries in multiple scenarios, higher energy/power densities and better cycle stability are pursued. Therefore, a possible solution may include continuously designing active electrode materials with high performances. Current solid electrolytes show inferior ion conductivities than conventional liquid electrolytes, inducing high internal resistances in batteries. Enhancing the electrochemical performance of flexible batteries can also be accomplished by designing solid electrolytes with ion diffusion channels and higher conductivity. The structural configuration defines the charge distribution and structural integrity. Therefore, to ensure excellent electrochemical performance, the structural configuration needs to be optimized.

11.3.3 Safety

Safety is another concern faced by flexible batteries. For flexible batteries based on organic electrolytes, the hazards of combustion and explosion at short circuiting impede their applications. Therefore, one possible solution is to design flame-retardant electrolytes without sacrificing electrochemical performances. Another solution is replacing the organic electrolyte with an aqueous electrolyte. However, some of the electrolytes used are corrosive and even poisonous, calling for intact sealing of the flexible batteries. Moreover, the working voltage of flexible batteries connected in series can exceed 36 V, which is higher than the safe voltage of the human body. Hence, designing proper wiring and connection of the electrodes is essential and challenging, which is seldom investigated and researched. For the actual application of flexible batteries, it is desired to explore the methods to improve the safety of flexible batteries.

11.3.4 Scale-Up Production

The scale-up production of flexible batteries is reckoned as the most urgent challenge, which must be carefully considered for the real application. As mentioned in previous chapters, a single flexible battery module can store low energy and power densities. The scale-up production of flexible batteries paves the way to connecting many flexible batteries in series and parallel, thus increasing the working voltage, energy, and power output. However, the available flexible batteries are generally lab based and fabricated by hand, rendering the scale-up production of flexible batteries difficult. Challenges mentioned above, such as improving the quality of flexible electrodes, designing high-performance solid electrolytes, eradicating hazards, and optimizing fabrication procedures, still hinder the mass production of flexible batteries. Moreover, for fiber flexible batteries, an advantage of compatibility with the current textile industry is recognized. However, the state-of-art fabrication of current fibrous batteries is still immature for operation in the industrial production line. Therefore, one of the most important future directions of scale-up production lies in finding an approach to combine the fabrication process of fiber batteries with the available production system.

In summary, since their initiation in the early 21st century, the past decades have witnessed the prospering of flexible batteries for smart electronics. Development has been made in designing high-performance component materials, efficient and robust battery structures, integration with other functioning parts, and scale-up production for future application. Hence, significant improvements have been made in flexibility, miniaturization, and wearability, revealing their competitiveness as power sources for future electronics. Despite the advance, the practical application of flexible batteries is still hindered by challenges such as imperfect flexible electrodes, comparatively inferior electrochemical properties, safety problems, and difficulties in scale-up production. Therefore, the pavement for the development of flexible batteries is clear. With the endless pursuit and efforts from both academy and industry, we sincerely believe that flexible batteries will eventually take an important role in our daily lives and make a difference soon.